COSMOLOGY 101

COSMOLOGY 101

David H Levy

ibooks

NEW YORK
www.ibooksinc.com

DISTRIBUTED BY SIMON & SCHUSTER

An Original Publication of ibooks, inc.

Pocket Books, a division of Simon & Schuster, Inc.
1230 Avenue of the Americas

Copyright © 2003 ibooks, inc.
Text copyright © 2003 David H. Levy

Parts of "Part Four: The Magic of Eclipses" previously appeared in *Eclipse, Voyage to Darkness and Light,* by David H. Levy, copyright © 2000 ibooks, inc.

An ibooks, inc. Book

All rights reserved, including the right to reproduce this book or portions thereof in any form whatsoever.
Distributed by Simon & Schuster, Inc.
1230 Avenue of the Americas, New York, NY 10020

ibooks, inc.
24 West 25th Street
New York, NY 10010

The ibooks World Wide Web Site Address is:
http://www.ibooksinc.com

ISBN 0-7434-5908-3
First ibooks, inc. printing June 2003

POCKET and colophon are registered trademarks of
Simon & Schuster, Inc.

Cover design by Mike Rivilis
Interior design by Gilda Hannah

Printed in the U.S.A.

IN MEMORY OF MY DADS

My father Nathaniel Levy
(March 27, 1909–February 3, 1985)

My father-in-law Leonard Wallach
(October 25, 1918–September 13, 2002)

Every star in the night sky, a friend once said, is a small "Yitgadal" in memory of a loved one. If that is so, then two stars must be twinkling extra strongly tonight, and I'll be watching for them.

ACKNOWLEDGMENTS

September 19, 2002: As I write this, I look out the window of an American Airlines 757 aircraft at Venus, hanging in the southwestern sky, a planet in space watching over the final words I write on my latest book. It is such a good feeling to finish a book, in the hope that its words and chapters will help its readers appreciate the cosmos. Therefore, my first acknowledgment definitely goes to the readers who have followed my writing in the past and who have made suggestions both for new books and for revisions of earlier books. This particular effort would not have been possible without the help of my wife, Wendee, who encouraged me in its conception and development, and assisted with my research, and who watched Venus with me from the airplane tonight. Sanford Wallach also helped with research. My publisher Byron Preiss offered his usual expert advice, and Howard Zimmerman did a fine job editing the manuscript.

May you enjoy learning about the cosmos from the pages to come, and may your journey through it be a pleasant one.

CONTENTS

| Preface | Of Eclipses, Cosmology, and History | 9 |

PART I: OF BIRTH, TIME AND SPACE — 13
- Chapter 1: More Things in Heaven and Earth — 14
- Chapter 2: With Restless Violence — 25

PART II: OF PLANETS AND SUNS — 37
- Chapter 3: Planets and Moons — 38
- Chapter 4: Planets of Distant Suns — 56
- Chapter 5: The Story of the Stars Above Us — 68
- Chapter 6: Variable Stars — 83
- Chapter 7: When Stars Die — 92
- Chapter 8: The Undiscover'd Country — 100

PART III: OF GALAXIES AND COSMOLOGY — 109
- Chapter 9: The Great Clusters — 110
- Chapter 10: The Milky Way — 119
- Chapter 11: Those Magnificent Galaxies — 138
- Chapter 12: Galactic Exotica — 149
- Chapter 13: A Modern Cosmology — 160

PART IV: THE MAGIC OF ECLIPSES — 177
- Chapter 14: Childhood Impressions of a Darkened Sun — 178
- Chapter 15: All About Eclipses — 183
- Chapter 16: Of Cycles and Friends — 189
- Chapter 17: Sun, Moon, and Surprise — 195
- Chapter 18: The Power of Gravity — 203
- Chapter 19: The Eclipse Experience — 212

PART V: TELESCOPES AND OBSERVING — 221
- Chapter 20: Telescopes — 222
- Chapter 21: Using Your Telescope — 238
- Chapter 22: Notes on Some Western U.S. Observatories — 245

PREFACE

Of Eclipses, Cosmology, and History

"These late eclipses in the Sun and Moon," the Earl of Gloucester complained in Shakespeare's *King Lear*, "portend no good to us." Shakespeare was an astute observer of "the phenomena of Creation" (Tracy, 1971) that we call Nature, and cosmology can be thought of as the ultimate expression of Nature. In its broadest sense, "cosmology" can be defined as the study of the totality of the phenomena in space and time of our Universe. By this definition, cosmology includes anything from the orbit of an electron around the nucleus of an atom to the motion of a galaxy through space. As a general term, cosmology is the astrophysical study of the history, structure, and dynamics of the Universe. A cosmologist would concentrate more on the Big Bang theory than on the motion of a planet around a star.

Our conception of cosmology, and its relation to ourselves, has changed much over the centuries as thinkers have tried to make sense of the vastness of space and time. Over the long history of human experience with the night skies, there was possibly no more crucial moment than the first decade of the seventeenth century, particularly the six-year period that began suddenly in the summer of 1604 with the appearance of a supernova. Added to the wonder of this sensational event was the series of eclipses of both the Sun and Moon that occurred in 1605. The series included a nearly total eclipse of the Moon the evening of April 3, and a partial lunar eclipse on the morning of September 27.

Here's more from the Earl of Gloucester in *King Lear* (act 1, scene 2):

> Though the wisdom of nature can reason it thus and thus, yet nature finds itself scourged by the sequent effects. Love cools, friendship falls off, brothers divide: in cities, mutinies; in countries, discord; in palaces, treason. . . .

After Gloucester exits, his bastard son Edmund disputes his words:

> This is the excellent foppery of the world, that when we are sick in fortune, often the surfeits of our own behavior, we make guilty of our disasters the sun, the moon, and stars, as if we were villains on necessity.

These lines are from the longest celestial reference in all of Shakespeare. It occurs at a crucial moment in the play, when the major part of the action is being set up; this is no afterthought. Could Shakespeare have been taking advantage of recent events when he composed this discussion about humanity and the cosmos in *King Lear*? Since Gloucester talks specifically about eclipses of both the Sun and the Moon, we should look for periods of time around the composition of the play that *pairs* of solar and lunar eclipses took place over London. This is a rare happening anywhere, but in early 17th century London it happened twice. In addition to the two eclipses in the fall of 1605, there was a pair of events in 1601, an almost total lunar eclipse on the evening of December 9 and a deep partial solar eclipse the day before Christmas. We do know from certain lines of evidence that *King Lear* could not have been written before 1603, and since James I saw it during the Christmas season of 1606, Shakespeare had to have written it by then. Shakespeare might well have had the three eclipses of 1605 in mind when he wrote Gloucester's lines.

Cosmology, Literature, and History

Even though I have always been interested in Shakespeare and in the night sky, it is only recently that it seemed reasonable for me to try to join the two subjects. All these years I accepted the conventional wisdom that Shakespeare, the world's best known writer, wasn't serious about the night sky, and that his astrological allusions amounted to the level of a love note from Hamlet to Ophelia (*Hamlet,* act 2, scene 2) that read:

> Doubt thou the stars are fire,
> Doubt that the sun doth move,
> Doubt truth to be a liar,
> But never doubt I love.

It wasn't until I really thought about the eclipse discussion in *King Lear* that I sensed something more was at work. The time during which the play was written was so critical—here was a window into history, an opportunity to look into the past and see how people in that distant time looked at the sky and discover what they thought about it. When we look at the night sky, we do not do it in isolation. We are part of a train of people stretching thousands of years back in time, who all share the common experience of looking up at the sky and wondering about what it contains, and what it means.

This book, *Cosmology 101*, offers an informal look at our opinions—of we who happen to be looking up in this era of time. But the next time we do look up, I call two stars in particular to your attention. One is Alpha Triangulum Australis, a southern star that is 415 light years away. The other is Albireo A, the brighter component of one of the most famous double stars in the northern sky. Albireo is 385 light years away. The light from the southern star left for us some 15 years before "these late eclipses" took place. Albireo's light departed the star some 15 years after them. We see these two stars as they were during the height of Shakespeare's lifetime, during a time when, as now, books about our understanding of the cosmos were being rewritten and Hamlet was saying "There are more things in heaven and earth, Horatio, than are dreamt of in your philosophy." (*Hamlet,* act 1, scene 5.)

High above our backyards, in the vastness of space and time, two distant suns are bringing us back to another time in Earth's history, when humanity was trying to understand the cosmos, and was about to discover that a new device, the telescope, would prove indeed that there are more things in heaven and earth than we could possibly dream of.

REFERENCES
Harrison, G. B. "These Late Eclipses." *Times Literary Supplement* [London] 30 Nov. 1933: 856.
Parr, Johnstone. *Tamburlaine's Malady, and Other Essays on Astrology in Elizabethan Drama.* University, AL.: University of Alabama P, 1953.
Tracy, Clarence. Lecture. 12 Oct. 1971.

PART ONE

CHAPTER 1

More Things in Heaven and Earth

Five billion years ago, a huge cloud of hydrogen gas swirled around in space. The cloud was thickest at its center, so dense that the gas glowed from the heat of its own gravity. At the outskirts of the cloud were other concentrations of gas and dust. As the cloud spun faster, its center grew hotter, until, at one single second in time, it ignited in a burst of nuclear fusion to begin the life of the Sun.

With that ignition, the remaining gas and dust in the clouds outlying regions congealed and accreted to build at least nine planetary worlds, several dozen smaller worlds, and many comets and asteroids. At least two of these worlds were about the same size, one-fourth that of Earth. One moved around the Sun in a lazy elliptical path. It still does, and we call it Mars. The other world flew round the Sun in an elongated orbit, alternately moving out into the fringes of the solar system and then tearing through the realm of the inner planets. Many times in that primordial history, this world came so close to Earth that its bulk would fill half the sky as it raced by, causing massive earthquakes and tidal disruptions before moving away. Finally, on one frightening day some four and a half billion years ago, this world came too close, sideswiping, moving away briefly, and then slamming into Earth.

The resulting explosion was so vast that it melted the entire crust of Earth. The other world broke apart, its pieces spinning out of control and colliding with each other and with Earth, again and again, as a ring of debris from both the Earth and the doomed world formed high above our planet. Over the next year, the ring's particles collided with one another, accreting to form at least one, and possibly two, new moons. One of them, too close, broke apart again, its pieces colliding with Earth in still more devastation. The other, farther away, formed the Moon we know and love.

At its creation the Moon was so close to Earth, a mere 10,000 miles away, that its immense tidal pull caused massive earthquakes, volcanic eruptions, and water tides on a daily basis. As time passed, the Moon crept slowly farther from Earth, and still does, at the rate of about a yard per century.

As it moved outward, the Moon's gravity affected the tides of Earth less and less. As it took up less space in the sky, the Moon seemed to shrink, gradually becoming the same apparent size as the Sun, which of course is much bigger but much farther away. And as the Moon continues to orbit the Earth, occasionally it passes in front of the Sun.

And that, in a nutshell, is why we have eclipses of the Sun.

Of the Moon, Time, and Space

Of all the observing sessions I have enjoyed, few match the experience I had at Acadia University on the hazy and foggy night of October 27, 1971. Carrying a small cassette recorder, I walked out on the ancient dykes that separate the verdant farmland from the tidal marshes near campus. The dyke system is very old; parts of it date back to the Acadians who built them before their expulsion in 1755. More than two centuries later, all was quiet as I walked slowly along the top of the dyke. I found a spot, sat down on the grass, looked up at the Moon, and started to listen to Beethoven's *Moonlight Sonata*. "It was an experience of peaceful magnificence," I wrote that night, "especially, when, in the middle of it, a fog enveloped almost everything but me and some stars. University Hall [the main building on campus] seemed to hang from the sky."

Nineteen months earlier, on March 7, 1970, I witnessed the partial phase of a solar eclipse in Nova Scotia, Canada. Although our group was in the path of totality, all we saw was a very dark lunar shadow projected against a layer of stratus clouds. The sky did clear later that day, however, so by the time I got back to my home at Acadia University, I had a beautiful view of the Annapolis Valley. It's a beautiful place—gentle hills outline the fields of apple trees, and the Annapolis-Cornwallis river winds its way, providing the valley with the ample irrigation it needs. But depending on time and tide, you will see one of two rivers, one gentle as fresh water rolls out into the bay, the other a raging torrent as seawater surges inland from the Bay of Fundy. At the height of the incoming tide, a flow of water greater than the combined current of all the rivers on Earth roars past Cape Split, just a few dozen miles away. This great pull on all the waters of Earth is caused mostly by the gravitational force of the Moon, aided in small part by the more distant Sun. If you watch the tides anywhere, you are watching an interplay of the forces of gravity among bodies in the Universe. By the time we returned to the valley a few hours after the total eclipse, the tides that resulted from the Sun and Moon being in the same spot in the sky were strong.

Years later, I returned to Nova Scotia for a visit, this time to study the tides more closely. At Cape Split these forces are as strong as they are anywhere in the world. If you stand at the Cape, you can hear the Moon roar.

The tidal marshes of Acadia are as good a place as any to begin a journey through space and time and to watch as "Time marshes on," as my wife Wendee says. We are going to board a ship of thought, leave Earth, and travel outward.

Fueling the Starship:
How Do We Measure Distances in Space?

As our journey begins, we'll measure distances first using the 93-million-mile distance between the Earth and the Sun—the yardstick known in science as an astronomical unit (AU). Farther out, we use a unit called a light year, the distance light and other forms of radiation travel in an Earth year, which is about six trillion miles. A beam of light could travel around the Earth seven times in a single second. But light still takes eight minutes to travel from the Sun to Earth, and more than four years to travel from the nearest star, Alpha Centauri, to Earth. Astronomers also use a different unit called a parsec, which is 3.3 light years.

The distance around the circumference of the night sky is measured in degrees—360 degrees goes around the sky. A degree has 60 arcminutes, and a minute has 60 arcseconds. If a star's position in the sky is measured at six-month intervals, when the Earth is on one side of the Sun and then at the other, it will appear to move slightly because of the change of position of the Earth. That motion is called parallax. A star that moves 10 arcseconds, for example, during that six-month period has a parallax of 10 seconds, and is said to be ten parsecs away. The distances to far galaxies are measured in thousands of parsecs, or megaparsecs. (Because it is a more difficult unit to understand than a light year, however, it will not be used generally in this book.)

Our starship is now fueled and, as we prepare for liftoff,

we take a last look outside and see the tidal basin filled to the top with salt water. All is well as we light the engines and lift off into space.

Our Solar System
Departing Earth, we race through the atmosphere and into space. We look back to see our beautiful blue planet receding in the window as we head toward our first stop, the Moon. Its surface holds the battle scars of ancient attacks by comets and asteroids. Its message to us, no doubt, is that our own home, the Earth, has also gone through a similar bombardment. The Moon is a much better keeper of her history: once a crater, always a crater. Earth's never-ending forces of mountain-building and erosion have erased most of its cratered evidence. Our Moon is about 240,000 miles away; light traveling from it takes about a third of a second to reach Earth.

As we pull away from the Earth-Moon system, we move toward Mars, a planet that might once have harbored simple forms of life. Could it have also been home to intelligent life? Such a feat would have been all but impossible for Mars; even if life managed to get a start on Mars at the same time it did on Earth, early in its history Mars lost all its water and its ability to sustain advanced forms of life.

We have little trouble navigating the asteroid belt; even though some 40,000 asteroids have been numbered and catalogued since 1800, there is a lot of space between these little rocky worlds. Our next stop is Jupiter, the king of the planets. We bid Jupiter thanks for being the cosmic vacuum cleaner—if Jupiter's gravity didn't change the orbits of so many comets, Earth would still be a sitting duck for major cosmic impacts every century. Jupiter still is showing some wounds, in the form of elevated levels of carbon monoxide, from its most recent comet encounter, with Shoemaker-Levy 9 in 1994.

We move on past Saturn's magnificent rings, Uranus tilting over on its side, and distant Neptune. Our next stop is Pluto and Charon, two icy worlds at the edge of the solar system. As we look back to our starting point, the Earth, with all its oceans, clouds, continents, and life, is but a faint speck of light. The Sun appears as only a bright star. Pluto gives us our first real opportunity to see the solar system as what it truly is—one small star surrounded by a few tiny worlds. The solar system is a very small part of our Milky Way Galaxy. We are now about 39 AUs from Nova Scotia.

We continue moving out, beyond the point where the Sun's outward-blowing wind of energy encounters the winds from the other stars. This is the heliopause, the true border of the solar system.

On to the Stars
We pass through the heliopause and move on through our galaxy. Compared to the solar system, our galaxy is unimaginably huge. It is shaped like a spiral pinwheel, and is so big that for the pinwheel to turn just once takes some 225 million years. Its 400 billion stars are varied in type, size, temperature, and age. Our Sun is one of the simplest of these stars. Its light is steady, and aside from a little too much ultraviolet, is just perfect for the kind of warmth that life needs on Earth. The closest star to the Sun is a system of three stars called Alpha Centauri. Here in Alpha Centauri's neighborhood, light from the Sun takes more than four years to reach us: We are four light years from Nova Scotia.

The Milky Way contains many kinds of stars. The most common is a red dwarf, a small and cool type of star that burns its nuclear fuel very slowly over many billions of years. We don't *see* too many such stars because they are so small and faint. The closest star to our Sun, a member of the Alpha Centauri system, is a red dwarf. More than half of our galaxy's stars are systems containing more than one sun. These two-star or multiple-star systems revolve around each other in periods ranging from a few hours to hundreds of years.

Our next stop: a big sun called Sirius. Eight light years from Earth, this star has a dense companion star called a white dwarf. How dense? A cubic inch of matter from it would weigh a ton on Earth! We move on to 47 Ursae Majoris, a faint star seen from Earth below the bowl of the Big Dipper. This star has a solar system consisting of at least two large planets. [Our current technology allows us to mostly identify large planets in orbit around distant suns. This doesn't mean these systems don't have smaller planets as well.]

We move on deeper into the Milky Way, past the Pleiades, a star cluster containing several hundred stars, all bound gravitationally to each other and moving through space together. The Pleiades is a beautiful grouping of hot blue stars, some surrounded by gas. Probably all stars, including our Sun, were born in clusters. The cluster of our Sun must have been a beautiful sight indeed, five billion years ago. Its stars have long since left their cosmic nest and have spread throughout the galaxy. We have no way of knowing which of the distant stars we see in the night sky are sisters of the Sun.

Through the Galaxy
Our ship sails onward, stopping by the great nebula in Orion. From Earth it is visible as a faint misty spot to the naked eye, and even the smallest telescope shows its beautiful strands of cosmic gas. But here in our starship, flying past its majestic swirls has got to be an unimaginable experience. We are 1,500 light years away from Nova Scotia. We next visit the Omega nebula, with its many infant suns, and the nearby Eagle nebula's exquisite pillars of darkness, each the home of a nascent solar system.

These nebulae contain young stars, stars just beginning

their own paths through the galaxy and solar life. We next visit stars like the Sun, but much older. These stars have used up their supplies of hydrogen fuel and have swollen and turned red. We drop by one of these, called Mira, and watch it slowly expand in size, then shrink. Its whole cycle lasts nine months. We see a Cepheid variable too, a bright blue star that varies not over months but in just a few days. Our next stop is by a large, fat star lazily revolving around a small, aggressive one. The small star spends its time robbing the larger star of its supply of hydrogen, and storing it in a surrounding disk. Unable to absorb the stolen material, the star undergoes a violent nuclear explosion every few months, then the process begins again. Another stop: a star that lost its life in a violent explosion during which the star shone briefly with the brightness of a hundred billion suns. All that's left of this supernova is a compact neutron star that spins hundreds of times per second.

We avoid stopping by a black hole, the ultimate end for stars that are much more massive than the Sun. After such a star explodes, it is unable to stop the process of shrinking to the size of Earth, the Moon, North America, Canada, Nova Scotia, the town of Wolfville, a city block, a house, a room, a book, and finally the period at the end of this sentence. The gravity is all still there, but it is concentrated into something the size of that period. Nothing—not even light—can escape from it. We rush past and head farther out to Messier 13, a globular cluster that travels at the very outskirts of the galaxy. Comprising hundreds of thousands of stars, these clusters are among the most beautiful objects in the galaxy. In the night sky as seen from Earth's southern hemisphere, Omega Centauri and 47 Tucanae are sparkling giant clusters. As we approach the cluster, we stop to refuel at a planet. Stepping out in the cool evening, we look up and see the mighty cluster taking up a quarter of the sky. What a brilliant sky this is, something hardly imagined back home! We are now 23,000 light years away from Nova Scotia.

The Galactic Neighborhood
As our journey continues outward, a big change can be seen as we begin to leave the galaxy. We first become aware of our exit as we see fewer and fewer stars around us. Looking back, the view is filled with the spinning spiral of the Milky Way.

Ahead are the distant spirals of the far-off galaxies. Since all stars belong to galaxies, the spaces between the galaxies appear empty and black. The shapes of the 24 galaxies that belong to our galactic family, called the Local Group, appear first. Two of them are the famous Clouds of Magellan, studied by the great explorer as he charted the southern seas on his voyage round the world. These galaxies are far from us—160,000 light years away—but they are relatively close to the Milky Way. It is likely that in the distant future the Milky Way will swallow the smaller one, making its stars part of itself.

The great galaxy of Andromeda, two million light years out, and our own Milky Way are the biggest in the Local Group. As we enter this new star system, we look back at our own galaxy, its 400 billion stars looking like a bright fuzzy spot against the background of space. If we spent enough time here, we might possibly find Earthlike planets teeming with life. But we must move on to Messier 33 in Triangulum. (This galaxy is No. 33 in the catalogue of Charles Messier, the famous 18th-century French comet hunter. He kept a list of objects that masquerade as comets.) Unlike Andromeda, which is tilted as seen from the Earth, the Triangulum Galaxy shows itself face-on. Scientists looking for signs of intelligent life in other galaxies have focused on this one, to see if some galactic civilization flourished there. Nothing has been found—yet.

MORE THINGS IN HEAVEN AND EARTH

Traveling at the Speed of Thought

As we move further out, our velocity continues to increase. By now we need to be moving at many times the speed of light, for to get to the next cluster of galaxies in the constellations of Virgo and Coma Berenices, we must travel some fifteen million light years from Nova Scotia. It is impossible to travel at the speed of light, but even if we could it would take us 15 million years to reach this nearest edge of the cluster. At the speed of *thought*, however, we can get there. Our Local Group is gravitationally bound to this cluster— thousands of galaxies sharing the same part of space, sharing the same destiny, all part of a big array called the Local Supercluster. Some of its galaxies are spirals like our Milky Way, others are irregular in shape, and still others are elliptical. Messier 87, one of the ellipticals, is also one of the largest galaxies in the observable universe. A big jet of light roars out from the center of this huge mass of stars. We are 70 million light years from Nova Scotia.

M87 is in the constellation of Virgo. We stay in Virgo as we go out not millions but billions of light years. 3C-273 is a quasi-stellar object, or quasar, the brilliantly energetic center of a distant galaxy. Because they are so bright, quasars are milestones, beacons to give us an idea of the enormous scale of the Universe. The most distant known quasar might be as far away as seven billion light years.

Recently quasars have offered us a key to understanding the scale of the Universe in a different way. The gravitational lens, predicted by Albert Einstein, is an example. The first observed incidence of gravitational lensing was the discovery of two identical quasars, almost touching each other in the sky. Careful study of the quasars revealed the presence of a very faint galaxy between them. The two quasars, it turns out, are really just one. On its way to Earth, the quasar's light encounters a galaxy, still very distant from us, but much closer than the quasar. The galaxy's massive gravitational field splits the light of the quasar into two, so the quasar appears as double. As we go farther out, we find a quasar split into four pieces by the same type of process. This marvelous sight is the Einstein cross. As we pass the intervening galaxy, we reach the single quasar much further out in space and back in time.

By turning big telescopes into areas where we know there are lots of galaxies, astronomers on Earth have seen even more exotic sights—distant quasars whose light is ripped apart into beautiful arcs of light. But on our starship trip, we see none of these gravitational lenses. As we go out in space, and further back in time, past galaxies, clusters of galaxies, and distant quasars, is it possible that our journey can take us back to the beginning of time itself, the explosion that started the Universe? And if it did, what would that moment look like? On that unanswered question . . .

Let's look at our watches. One shows the time in our spaceship. Since we are moving much faster than the speed of light, our watch is running backwards, almost to the start of time. The other watch shows the time in Nova Scotia. In this journey only three hours have passed. We turn around and start the voyage home. For another three hours, we rush past quasars, galaxies, the Large Magellanic Cloud, Omega Centauri, Alpha Centauri, and as we head northward into the solar system, we spot the orbs of the giant planets, and finally the little dot that grows and grows until it assumes the appearance of Earth. We come to a stop on the tidal flats in Nova Scotia, disembarking into a land utterly changed. Six hours ago the basin was full, but while we were away the gravity of the Moon did its work, and now the water is gone. After our 6-hour cosmic voyage, the landscape on our return is a vast brown sea of mud.

CHAPTER 2

With Restless Violence

"With Restless Violence round about the Pendant World"
—William Shakespeare,
Measure for Measure, act 3, scene 1

In a cosmic sense, the collision of the ninth periodic comet discovered by the team of Carolyn and Gene Shoemaker and David Levy with the planet Jupiter was unremarkable. The history of the solar system, indeed its very genesis, has been marked by countless such events. . . . In human terms, on the other hand, the impact of Comet Shoemaker-Levy 9's 20-odd pieces with Jupiter was an unprecedented event of global significance. . . . For a week in July, the world looked up from its normal preoccupations long enough to notice, and to ponder, the awesome beauty of the natural world and the surprising unpredictability of the universe.
—Keith Noll, Harold Weaver, and Paul Feldman, 1996

On a clear night I often look at the sky and imagine myself in a conversation with Earth, asking it how many gorgeous nights it has seen like this one. "There've been nights," my partner in imaginary conversation answers, "where I have seen comets brighter than the Moon, getting brighter by the hour until they collide." In the long cosmic perspective of our planet, a peaceful night might be the norm, but violent nights are locked in her memory. Like a child unwilling to remember sad events, Earth tends to erode the evidence of these blasts, but a simple look at the Moon through a small telescope reveals hundreds of craters, each the remains of a cosmic impact there. And if it happened there, it happened here. The objects that caused these impacts, comets and asteroids, are still out there, visitors to our sky. Once every decade or so, a bright comet winds its way through our vicinity. Although there are trillions of comets, only about 1,500 have been observed passing through the inner regions of our solar system in the last three thousand years.

Formed from cosmic ices and dust over four and a half billion years ago, a typical comet journeyed about the solar system, as far from the Sun as Pluto is, for many hundreds of millions of years. Then some pull of gravity, perhaps from Neptune or from a passing star, changed its course. It started to head into the inner part of the solar system, into the realm of the planets. At one crucial time in the eons of its journey, the small village-sized world passed close by Jupiter and its orbit changed again. Halley's Comet might actually have made several close passages by Jupiter before it took up its present path, which brings it into the vicinity of the Sun once every 76 years, once every average human lifetime.

How many ancient peoples of Earth must have seen the wonder of a comet in the night? It was common in those times to see comets as omens of terrible events, and therefore observers kept careful records of their appearances in the sky. The oldest known comet appeared during a war between two Chinese kings, Wu-Wang and Chou, around 1059 B.C. Then in 240 B.C., Chinese records show that a comet appeared that has now been identified as the earliest known appearance of Halley's Comet. It managed to appear at other crucial times in Earth's history, earning a place on the famous old Bayeux Tapestry. This long, beautiful embroidery was completed in 1083 and still is displayed in a church in Bayeux, France. The tapestry depicts the victory of William the Conqueror over the newly crowned King Harold of England at the Battle of Hastings in 1066. *Istimirant Stella*, the inscribed words describe the terrified population: They are in awe of the star. In 1456 the comet's appearance terrified Pope Calixtus III. If legend is true, the Pope excommunicated the comet, calling it a devil's device. The Pope did establish the midday *angelus* prayer at that time to help protect against the comet's wrath. In 1531 the comet might have been glimpsed by an Englishman named Richard Shakespeare and his small child John; in 1607 it was possibly viewed by John's son William. The comet appeared a few years after William ascribed these lines of his famous play to Calpurnia, wife of Julius Caesar:

> When Beggars die, there are no comets seen,
> The Heavens themselves blaze forth the death
>   of princes.
> —*Julius Caesar*, act 2, scene 2

When the comet returned in 1682, one of its many observers was the young scientist Edmond Halley. Halley did the calculations that led him to suspect that the comet of 1682 was the same as the comets of 1531 and 1607. When Johann Georg Palitzsch, a German farmer, found the comet on Christmas night, 1758, humanity's understanding of the Universe would forever change.

The story of the taming of Halley's Comet is one of the most important in the history of science. Instead of heaven-sent terrorizing apparitions in the night, comets were now seen as members of the family of the Sun, part of the uni-

Halley's Comet: June 18, 1986. I took several images of Halley's 1985–86 visit to the inner part of our solar system. Using a 20-inch telescope, Steve Larson of the University of Arizona and I took a number of short exposures and then combined them in a process called co-adding to achieve this image of the comet heading in our direction.

verse just like Earth and the other worlds, visiting us not to foretell great events but because they are merely following their orbits about the Sun. Edmond Halley's contribution went far beyond comets. He took the night sky out of the realm of local superstition, and brought it to a level of the Universe, where worlds obey physical laws.

A Comet Primer

Halley's comet has visited several times since then. In 1835 it witnessed the expansion of the United States, and the birth of Samuel Clemens (a.k.a. Mark Twain). In 1910, true to his prediction, the comet saw his demise. As the comet neared Earth in March of 1986, an armada of spacecraft met it and studied its composition to learn that the comet was made of many organic particles containing carbon, hydrogen, oxygen, and nitrogen. Now known as CHON particles, these particles are the alphabet-soup of life. The comet was also confirmed to have a substantial supply of water.

When a comet is far away from the Sun, it resembles a very large snowball, perhaps a few miles across. As a comet approaches the Sun, its ices turn to gases in the process of sublimation, which allows the comet's dust to escape as well. As the gentle pressure of the Sun's radiation—the solar wind—acts on the comet's particles, they stream away from it to form tails of gas and of dust.

Each comet is unique, with an appearance different from all others. The ancient Chinese tried to divide comets into two broad categories—the *po*, or bushy star comet (with a large fuzzy coma, usually without a tail) and *hui*, or broom star comet, with a tail. Comets have different appearances for a variety of reasons. Comets differ intrinsically; they have varying sizes, different proportions of gas to dust, and differing amounts of crust to prevent or allow the outflow of gas and dust. They are also different geometrically; the angle at which we see a comet from Earth can result in different appearances for the same comet. Comet Halley in 1910 covered much of the sky, but in 1986, viewed from a greater distance and different angle and distance from Earth, the same comet was far less spectacular.

Meteors: Dust Particles from Comets
Most of us have not seen a comet, but we've all seen a meteor streak across the night. A meteor is simply an event that takes place when a particle of dust, expelled long ago by a comet, encounters the Earth's upper atmosphere at a high velocity. Crashing in at up to 40 miles per second, the particle, called a meteoroid, heats the air around it so that it glows, and it is that glowing that we call a meteor. These particles travel around the Sun, and no doubt other stars, in swarms that essentially are the same as the orbit of the comet. For example, every August 11 the Earth plows through the swarm of debris left by Comet Swift-Tuttle, a comet discovered in 1862 that returned in 1992. The swarm lies all along the comet's orbit, so we get a meteor shower every summer.

What happens if the Earth passes through the comet orbit shortly after the comet itself has been there? In the early morning of November 19, 2001, the Earth moved inexorably through a very dense swarm of particles from Comet Tempel-Tuttle. For a time, meteors appeared at the rate of more than one per second. All the meteors come from the same direction, traveling along parallel paths. But because of perspective, the meteors that night seemed to come out of a single point in the sky, in the constellation of Leo the Lion. The effect is like looking down a railroad track. Even though the rails have to be exactly parallel, they appear to converge in the distance.

Asteroids, Comets, and Global Devastation.
On the calm, clear morning of June 30, 1908, the air was shattered by a huge explosion as something exploded above the

Tunguska River in Siberia. This was no mere particle the size of a sand grain. The object that collided with Earth that day was an asteroid the size of a small building.

When the first asteroid was discovered in 1800, astronomers found that its orbit lay entirely between the orbits of Mars and Jupiter. Within only a few years, several other asteroids were found, all between the two planets. Until 1931, astronomers believed that *all* asteroids traveled in the space between the orbits of Mars and Jupiter. That year an asteroid now named Apollo passed close to the Earth. A few years later an asteroid bright enough to be seen without a telescope rushed past the Earth. Now known as Hermes, it has not been seen again, but it will be some night in the future as it plays a close game of tag with Earth once more. We have no idea when, or how close, Hermes's next visit will be. It could pass as close as it did in 1937, and someday, probably in the distant future, it may strike the Earth.

While thousands of small particles hit the Earth each day, something the size of a building should encounter us not more often than once a century. The larger the objects are, the less likely their chances of colliding with Earth. Asteroids the size of Hermes and Apollo probably hit the Earth about once every hundred thousand years, with devastating results. Clouds of dust would form after an impact and quickly spread until all Earth would be covered by a cloud dense enough to prevent any sunlight from coming through for months. The loss of sunlight, and photosynthesis, would result in worldwide crop failures and starvation on a massive scale.

Although impacts of this strength would be disastrous, they pale beside the catastrophic strikes of asteroids or comets that are several miles across. Such events are strong enough to cause mass extinctions of large numbers of species. The most infamous of these occurred 65 million years ago. Although we are not sure if the intruding object was a comet or an asteroid, we have some clues from the population of objects that orbits the Sun today. There are only one or two asteroids, close to the necessary size, which do travel in orbits that cross that of Earth. However, large comets approach Earth quite frequently: In 1983, Comet IRAS-Araki-Allcock came to within four million miles, a stone's throw in space.

Something that size, either a comet or an asteroid, crashed into the Earth sixty-five million years ago. If the impactor was an asteroid, it crept up on Earth surreptitiously until it crashed through the atmosphere. But if it was a comet, the dinosaurs saw it brightening in the days and hours before the crash. The comet plowed through Earth's atmosphere, destroying the fragile ozone layer before striking the ground just off the coast of modern day Yucatan, carving a crater more than a hundred miles wide. Debris pouring out of the crater landed in places hundreds of miles away, and finer dust quickly spread around the world. As temperatures rose to the level of an oven set to broiling, a good proportion of Earth's biosphere was incinerated as ground fires ignited all over the world. Mile-high tsunamis flooded coastlines all around the Gulf of Mexico, and traveled around the world inflicting more damage to low-lying areas. Meantime, the upper atmosphere was rapidly filling with a planetwide dust cloud that, at its thickest, meant that the planet did not receive any sunlight whatsoever. Global temperatures plummeted as all outdoors became black as a darkroom. After many months the dust cleared to a hazy sunlight, and the temperature rose again with the onset of a worldwide greenhouse effect that lasted for centuries.

The big dinosaurs probably died out very quickly, possibly within a few weeks or months of the impact. The smaller, more agile creatures suffered under the impossible conditions that fate had brought upon them. By the time it was over, some 70 percent of all species of life had vanished from the Earth.

The evidence for this doomsday scenario lies in the pages

Comet Hyakutake. Arizona amateur photographer Dean Ketelsen took this magnificent photo as the comet sped past Earth in March 1996.

of rock that make up our planet's history book. In 1979, Luis and Walter Alvarez announced their discovery of a thin layer of clay, enriched with the precious metal iridium, at the boundary of the Cretaceous and Tertiary periods of Earth's history. Then, in 1991, the existence of a buried impact crater beneath the village of Chicxulub in Mexico's Yucatan peninsula was confirmed. The small Central American nation of Belize also contains debris from what must have been a monstrous tsunami.

Shoemaker-Levy 9 and the Origin of Life
No comet has ever been studied with more suspense than the small, fragmented comet called Shoemaker-Levy 9. This comet was important not for what it was, but for what it did: By colliding with Jupiter in the most spectacular explosions ever seen within the solar system, this comet provided a lesson about the origin of life.

I have already said that every comet is unique, but Shoemaker-Levy 9 took uniqueness to a new level. Even on the night of its discovery by Carolyn and Gene Shoemaker and me, its appearance was totally unlike anything seen before. Near the center of two pictures I took on the night of March 23, 1993, Carolyn found a "squashed comet"—instead of a coma and tail, there was a bar of fuzzy light, with several tails pointing northward. The comet, it turned out, had grazed by Jupiter just a few months before discovery. Just as the Moon pulls on the waters of Earth, Jupiter's gravity stretched this comet so strongly that it fell apart into many fragments, a "string of pearls," each pearl with its own tail.

The comet's strange appearance was only the herald of what was to come. Instead of orbiting the Sun directly, as every other known comet has done, Shoemaker-Levy 9 had orbited Jupiter since about 1929. On May 22, 1993, the International Astronomical Union issued a major announcement: Comet Shoemaker-Levy 9 was in its final orbit, and would collide with Jupiter during July of 1994. That gave the astronomical community an ample amount of time to plan—fourteen months—and the result was the biggest observational campaign for a single event put together in the history of astronomy. By the night of the first impact on July 16, 1994, virtually every major telescope on Earth was pointed toward Jupiter. In space, the *Galileo* spacecraft en route to Jupiter, and the Hubble Space Telescope, watched as the first fragment of Shoemaker-Levy 9, traveling at 134,000 miles per hour, tore into Jupiter's atmosphere. By the end of crash week, Jupiter's southern hemisphere was blackened by soot-like clouds. In the war of the worlds, the tiny comet obviously lost, but Jupiter was injured. Thick clouds, lasting for months, blackened parts of its southern hemisphere.

A Week to Remember

Before July 16, 1994, much of what we knew about comet impacts was theoretical. On that day comet impacts entered the world of reality when the first piece of Shoemaker-Levy 9, a fragment called "A," forced its way into Jupiter's upper atmosphere at the astonishing speed of 135,000 miles per hour. A bright fireball surged upward some 2000 miles above Jupiter's cloud tops.

On Monday, July 18, a tiny fragment approached Jupiter. It small flash was seen by at least one telescope. Thirty seconds later, the gates of hell appeared to open as fragment "G" blew up over Jupiter. By the evening of July 18, the "G" impact site was so dark that virtually anyone could see it through almost any telescope. Larger than the Earth, the impact sites from comet fragments "G," "H," "K," and "L" were clearly visible even to children looking through small telescopes. These were the most obvious features seen on Jupiter since Galileo first turned his telescope to the giant planet in 1610.

Lessons Learned

In witnessing the collision of Shoemaker-Levy 9, we also watched the birth of a whole new science. For one special week in July of '94, Nature let us peer into one of her most closely guarded secrets. Comet crashes and life have gone hand in hand in the long history of the solar system, and that includes Earth, from the early times as life was gaining a foothold, to the extinction of the dinosaurs 65 million years ago.

We have seen how at least one impact, which ended the age of dinosaurs, helped to shape the course of life on Earth. In fact, early in the Earth's history comets might have done more than that, by giving the Earth its supply of organic materials like carbon, hydrogen, nitrogen, and oxygen, and consequently its water supply as well.

On many clear, peaceful nights, I often think about the Earth's long history of contacts with other objects in space. Shoemaker-Levy 9 brought this lesson home, and added to it a warning about our future. In its aftermath, underscored by the dark impact points across Jupiter's southern hemisphere, there was a prediction that someday Earth could suffer the same fate. We learned that lesson well. During impact week in 1994, the U.S. Congress created the Shoemaker commission to find ways of discovering every potential threat in space a kilometer wide or greater. If we are vigilant and persistent, we might discover a killer comet or asteroid in time to turn a catastrophe into a near miss. If we are successful, Earth and her people would not have to worry about a violent night again.

REFERENCES

Alvarez, Luis W., et al. "Extraterrestrial cause for the Cretaceous-Tertiary Extinction." Science 208 (1980): 1095-108.

(MarsdenBrianIAU Circular) Marsden, Brian. IAU Circular. 5800, 5801, May 22, 1993.

Keith S. Noll, Harold A. Weaver, and Paul D. Feldman, *The Collision of Comet Shoemaker-Levy 9 and Jupiter* (Cambridge, England: Cambridge University Press, 1996), xiii.

PART TWO

OF PLANETS AND SUNS

CHAPTER 3

Planets and Moons

> Flamsteed, at the end of the last century, and Mayer and Le Moniuer, in this, had observed Uranus as a small star. But it was not until 1781 that Dr. Herschel discovered its motion, and soon after, by following this star carefully, it has been ascertained to be a true planet.
> —James Ryan, 1831

In 1850 Alfred, Lord Tennyson, published *In Memoriam*, the masterwork of his life. Far more than an elegy to a friend, this poem was a discussion of the latest ideas about the origin of the solar system, from the geological history of the Earth ". . . in tracts of fluent heat began . . ." to the evolution of humanity. A revolution of scientific thought was taking place at this time, and Tennyson's great poem tried to bring together new ideas about the formation of the solar system through Darwin's *Origin of Species*.

In 1846, as Tennyson was completing his elegy, a new

world, Neptune, was found at the edge of the solar system. Its story began on March 13, 1781 when William Herschel used a small telescope from his backyard to discover a new world that was subsequently named Uranus. Its orbit was quickly calculated and refined over time as more and more precise positions became available. By the 1830s, however, astronomers realized that Uranus was deviating from the orbit they had established for it. In 1841 John Couch Adams, a 23-year-old Cambridge University student, thought he had figured it out: Uranus's strange orbit could be explained if another planet was orbiting further out in the solar system, a world large enough that its gravity was affecting the path of Uranus. Just after he graduated from Cambridge, Adams sent his calculations of where this mighty new planet might be to his astronomy professor, John Challis, who in turn handed the work to George Airy, England's Astronomer Royal.

Although Airy was interested at a theoretical level, he did nothing about ordering an actual telescopic search. Then at the end of 1845, Urbain Jean Joseph Leverrier, a young French astronomer, also presented Airy with an orbit calculation and predictions for where the new world might be. Airy studied this paper and commented that Leverrier's predictions agreed within one degree of what Adams had predicted. Airy finally suggested that Challis search for the object. But two things went wrong: Airy forgot, or otherwise neglected, to tell Adams about Leverrier's work, and Challis, instead of going right to the predicted position, mounted a cumbersome star by star search over a large area of sky—Challis actually went over the new planet twice without recognizing it. Meanwhile Leverrier had no luck in getting a search going at the Paris Observatory in his native France, so he went to Johann Galle at Germany's Berlin Observatory. An intrigued Galle began a search at Leverrier's predicted position the next night, and found the new planet with ease.

After the discovery was announced, Airy tried to have Adams credited with it as well. The French were livid with Airy's belated attempt to get some English credit. As the French and English confronted each other, Adams met Leverrier, and the two became close friends.

In 1929, a young amateur astronomer from Kansas arrived at Lowell Observatory in Flagstaff, Arizona. His new boss, Observatory Director V.M. Slipher, told him that his main project would be to carry on the search for a new planet that Percival Lowell had started in 1905. Three searches had already been unsuccessful at the observatory, but the senior staff thought that their new, 13-inch diameter widefield telescope and a blink comparator might make the project more fruitful. Clyde Tombaugh learned that he could examine two photographic plates of the same part of the sky and tell quickly if any object had moved between the time the two plates were taken. He also expected that a trans-Neptunian planet should have a specific rate and direction of motion.

On February 18, 1930, Tombaugh was "blinking" two plates centered on the bright star Delta Geminorum. Just after four o'clock that snowy afternoon, he saw the two images of a faint object. "That's it!" he whispered. After checking the image with a third plate he had taken, he walked down the corridor to the director's office. "Dr. Slipher," he said, "I have found your planet X."

A Revolution

This story of how the solar system's outermost planets were found is one of the most exciting in astronomy. It certainly enthralled me when my father recounted it over dinner one evening during the summer of 1960, and I hadn't forgotten it 42 years later, when warm spring evenings in June 2002 brought me outdoors to see the major planets clustered together in the western sky. Mercury, Venus, Mars, Jupiter, and Saturn were all visible for the better part of an hour after sunset. These were the wandering stars of ancient peoples. They had a healthy way of looking at the solar system in terms of the entire sky, different from how astronomy is studied in today's colleges where the solar system is partitioned off from the rest of the Universe. To ancient cultures the sky was a unit. Beyond the Sun and Moon, the stars were perceived as a fixed sphere, and each of the five planets represented closer spheres. From the time of the ancient Chaldeans five thousand years ago, people of many cultures felt a need to record the changing positions and magnitudes of these bodies, partly out of a belief that these wandering stars might affect our lives. (The Greeks named these wanderers "planets" for the simple reason that planet *means* wanderer.)

On the last day of his life, in 1543 Nicholas Copernicus published his masterpiece, a work called *De Revolutionibus Orbium Coelestium* (*The Revolution of the Celestial Orbs*). In it, Copernicus proposed that Earth was a planet like the others, all orbiting the Sun. Death would spare Copernicus from the battle that would result from his theory. Two completely different episodes triggered that conflict, one involving a distant star, and the other, two pieces of glass.

Danish astronomer Tycho Brahe's discovery of a brilliant new star in Cassiopeia in 1572 took the wind out of the long accepted idea that the outermost sphere of the sky, containing all the stars, was unchangeable. (Brahe called it a *nova*— a new star. Today we know that it was a supernova—the cataclysmic explosion of a massive star.)

As the first such appearance since 1054, it really captivated the scientific world.

The second episode began early in 1610, when Galileo Galilei discovered that Jupiter had four moons that clearly revolved about it. Later that year, he studied the changing phases of Venus, and could find no explanation for those phases unless Venus and Earth revolved about the Sun in separate orbits. To anyone who looked through Galileo's telescope, it was obvious that the idea that the Earth was the center of the universe, and that everything must revolve about it, was no longer viable. With Galileo's new telescope, one could see the moons circling Jupiter through just a few nights of observation. It was the start of a long and insightful journey to the understanding of the neighbor worlds in our solar system.

This new understanding, however, came at a tremendous cost to Galileo. In 1611, just one year after Galileo's discovery of the moons, the Catholic Church filed the first of several documents against him—in secret and apparently without his knowledge. In 1616 a document known as the *Codex* forbade any teaching that the Sun is in the center of heaven. Galileo decided to wait until a more propitious time to defend the new idea. That time came, Galileo thought, when a new Pope was elected in 1623. Urban VIII seemed open to new concepts, and in the new Pope's first years Galileo walked and talked with him. But Galileo seriously misread these discussions. In 1632 he published his *Dialogue on the Great World Systems*, in which he presented both the Sun-centered and the Earth-centered universes as a debate. Unfortunately, the character in his book who represented the Earth-centered system was named Simplicio, and he seemed to mock the Pope. Urban VIII was outraged. He promptly put the old and virtually blind astronomer at the mercy of the Holy Office of the Inquisition. Galileo was forced to see the instruments of torture, and at a trial in 1632, he was forced to recant his observations and conclusions about the Sun being the center of the universe. He was also sentenced to spend the remainder of his life under house arrest.

Galileo would have been delighted that his work, and his suffering, eventually paid such dividends. Thanks to telescopes based on Earth and in space, we have a completely new understanding of our planetary system.

Mercury: World of Heat
The hardest of the major planets to see, Mercury is always close to the Sun. For periods not much longer than two weeks at a time, as Mercury goes through its 88-day orbit of the Sun it is visible after sunset or before sunrise in the twilight sky. The planet rotates once every 59 days, a day that is more than half as long as its year. If you stood on Mercury, you would see the Sun's appearing to move west, then east, in a strange path across Mercury's sky, staying above the horizon for 90 days at a time.

Since the planet does not have an atmosphere to temper the Sun's heat, the temperature on Mercury's sunny side tops 800 degrees F. On its night side, however, the temperature plunges down to some 300 degrees below zero. In 1974 NASA's *Mariner 10* spacecraft took clear pictures that showed Mercury's surface as riddled with craters as our Moon is. As we have seen earlier, most of these craters probably date from the period 3.9 billion years ago known as the Late Heavy Bombardment. During this time Mercury, Venus, Earth, the Moon, and Mars were all getting carpet-bombed by a heavy stream of comets and asteroids. If we were living at that time, we would see a night sky rich with bright comets, and there would be a distinct possibility of a major impact during our lifetime. Because Mercury and the Moon have no atmospheres to erode the impact evidence, the ancient craters still dominate those worlds.

Venus: A Clouded World
Our first indication that Venus's temperatures were as hot as Mercury's came in 1962, when the American spacecraft *Mariner 2* sailed past the cloud-veiled planet and found a surface temperature at near 900 degrees F. The atmosphere

is so hot because Venus is the result of a greenhouse effect run amok on a planetary scale.

For the first billion years of the history of the solar system, Venus, Earth, and Mars had much in common. Their three atmospheres were rich in carbon dioxide, and all three might even have had liquid oceans. Time has changed all that; the three planets now have very different atmospheres. Mars's atmosphere is very thin, and Venus's air is incredibly thick. Venus receives about twice as much sunlight as Earth. As temperatures soared in Venus's past, its oceans evaporated, increasing the amount of water vapor in the atmosphere. That water vapor, which has also vanished, helped increase the amount of carbon dioxide still present in Venus's atmosphere.

In 1991 the *Magellan* spacecraft completed the first of a sequence of maps of the surface of Venus, revealing details of a continentlike feature we call Aphrodite Terra that crosses the planet's equator and is marked by geologic faults. A second "continent," Ishtar Terra, straddles the planet's north polar region. *Magellan* also observed some craters, the results of impacts of asteroids and comets, but far fewer in number than on any other planet in the inner solar system.

Earth: A Delicate Home
After visiting Venus, we should appreciate what we have here on Earth: an atmosphere with 78 percent carbon dioxide instead of almost 100 percent, and enough oxygen to support a great variety of life. Unlike Venus, which is far too hot, or Mars, which is unbearably cold and without sufficient atmosphere, Earth seems just about perfect for life.

Instead of having a planetwide greenhouse effect like that of Venus, Earth's atmopshere is almost perfectly balanced. It is far enough away from the Sun that the solar heating is not nearly as bad as on Venus. A sufficient amount of heat that built up during the day is allowed to escape off into space at night. Despite the increase of chloroflurocarbons (CFC's) from our industrialized society, our greenhouse effect has not yet built itself into the catastrophic state that exists on Venus.

Earth has several "spheres." Stretching out from its central core are the mantle and the crust, and above them, the atmosphere is divided among the troposphere, the stratosphere, and the ionosphere. The narrowest and most delicate is the biosphere, the domain of all life that resides in land, sea, and sky. We know of no other world, in or out of our solar system, which has a biosphere.

The Earth in Space
Just by looking up, can we tell that the world we live on is traveling through space? There are actually several lines of evidence.

1. If you look toward the pole star and watch the stars around it each hour, it will soon become evident that they appear to move around the pole once a day. This is evidence that we live on a ball in space that completes a rotation in that time.

2. Stars rise about four minutes earlier each night, coming back to the same rising time a year later. This is evidence that the Earth is revolving about the Sun each year.

3. Meteors falling in the night incinerate in the atmosphere as Earth pushes its way through space.

Earth's Moon
When our planet was very young, its surface barely hardened, a large planetesimal world the size of Mars sideswiped the Earth. Earth's crust instantly melted, and material from Earth and the destroyed other world blasted out, collecting together in orbit about the Earth. Within a year, the whole mass formed a primordial moon. We've already seen how, more than 500 million years later (about 3.9 billion years

ago), a series of large objects struck the Moon during the Late Heavy Bombardment, gouging out the great impact basins that we see today as the dark regions some call the Man in the Moon. These basins no longer exist as the huge craters they once were. Some time after they were formed, great eruptions beneath the Moon's surface filled them with dark lava, giving them their dusky appearance.

Mars: Planet of Mystery

The red planet Mars revolves about the Sun in a bit less than two Earth years, and its day is only 40 minutes longer than Earth's day. During the summer of 1997 it hosted an American spacecraft called *Mars Pathfinder* as it recorded summertime highs of 15 degrees F, and nighttime lows to minus 110 degrees F. More recently, *Mars Global Surveyor* did a further study of the Martian surface, and in the fall of 2001 *Mars Odyssey* began its mission of mapping from orbit the locations and abundances of chemical elements and minerals, looking for water, and studying the radiation environment around the planet. Mars is a waterless world today, but billions of years ago it probably had it in abundance. In fact, *Pathfinder* landed amidst the remains of what appears to have been an ancient flood. Over time, Mars's water evaporated into space. As carbon dioxide increased in Venus's air, Mars lost its supply; what CO_2 is left resides in the rocky soil beneath the planet's surface.

At its most favorable apparitions, Mars can be as close as 35 million miles from us and twice as bright as Sirius, the sky's brightest star. Mars has a desertlike environment and a thin atmosphere that makes its sky pink. There are some spectacular mountains, especially Olympus Mons—a volcano the size of Arizona. On Earth, even large volcanoes like the Hawaiian Islands do not exceed a certain size because the crustal plates beneath them keep shifting, and old volcanic vents get shut off as new ones open. Since Mars apparently does not experience such shifting of its crust, volcanoes there can grow indefinitely.

Late in the 19th century, the Italian astronomer Giovanni Schiaparelli, a first-class observer, saw long, straight lines that seemed to crisscross the Martian surface. He described these lines as "canali" or channels. The American amateur astronomer Percival Lowell loosely translated that to canals, and his observations persuaded him that these channels were artificial constructs of a global Martian civilization struggling to make the best use of a dwindling water supply. These ideas persisted as late as the summer of 1965, when the *Mariner 4* spacecraft photographed a crater-strewn planet with no evidence of any life or canals. It seems that the channels are a result of observer bias, in which the eye tries to connect short linear features to form a longer one.

In 1984 a meteorite was discovered in Antarctica that was later determined to have begun its journey from Mars. A team of NASA scientists studying that meteorite suggested in 1996 that it could contain the remains of simple, submicroscopic life forms. This contentious finding is still under intense scientific scrutiny. Although the question of intelligent civilizations on Mars is now pretty much assured to be answered as negative, the question of any primitive life there is still open, and will probably remain that way until a spacecraft can return samples from this mysterious world.

Martian Satellites

In 1726, Jonathan Swift wrote in *Gulliver's Travels* that Mars was accompanied by two moons. What a prescient piece of fiction this was, published a century and a half before Asaph Hall, observing from Washington's U.S. Naval Observatory, actually found the two moons that became known as Phobos and Deimos: "fear" and "terror," respectively. These moons are possibly small captured asteroids. But if they are, they are not the only asteroids whose destinies were changed by Mars.

Early in this century, Max Wolf, an astronomer from Heidelberg, found the first of a group of asteroids so distant that they apparently "shared" Jupiter's orbit, circling the Sun some 60 degrees ahead or behind the giant planet along the curve of Jupiter's orbit. Known as Trojan asteroids, they oscillate about the Lagrangian points ahead or behind Jupiter. They are named after heroes in the Trojan War. Those on one side of Jupiter are named for the Greek heroes, while those on the other side are named for Trojan heroes, with a spy from each side in the other's camp. In June of 1990, astrogeologist Henry Holt and I found something completely new—a "Martian Trojan," an asteroid some two kilometers wide in a Trojan orbit with respect to Mars. Now called 5261 Eureka, the name recognizes the excitement Archimedes felt after making an important discovery. When asked to tell the difference between a crown made of real gold and a fake, Archimedes solved the problem while taking a bath and thinking up his principle that objects of different densities displace different amounts of water. So thrilled was Archimedes, the story goes, that he tore out of his bath and ran out of the house, presumably still unclad, yelling "Eureka! Eureka!" ("I found it!") We hope that other "Martian Trojans" will get names that express similar joy in making a discovery. The lucky find of a Martian Trojan excited the community of planetary scientists. There might have been Trojans around Mercury or Venus at one time, but there do not appear to be any now.

Jupiter: King of Worlds

Not only is Jupiter the mightiest planet, but it also rotates faster than the others—once in under ten hours! Jupiter is the closest representative of a very different kind of planet from the Earth. As with Saturn, Uranus, and Neptune, the planet is a giant ball of gas, an unappetizing mixture of hydrogen, helium, methane, and ammonia. Below the tops of the clouds might be a concentration of cold gases as thick as gelatin. Had it possessed many more times its mass, Jupiter might have started nuclear fusion and lit up as a star. Even at its size, Jupiter gives off four times as much radiation as it receives from the Sun.

Thirty thousand miles long and ten thousand wide, Jupiter's Great Red Spot dominates the planet's southern hemisphere. Some three times larger than the entire Earth, this spot has been observed since the mid-1600s, as long as we have been observing Jupiter with telescopes large enough to see it. This monster is a long-lived storm whose origin is a mystery. The dynamics of Jupiter's atmosphere are so complex that a storm this size could start spontaneously. It is also possible that the storm began as a result of a comet whose impact upset the atmosphere enough to start the cyclonic activity.

Jupiter's Moons
When Galileo first spied Jupiter's four largest satellites, he could have had no idea how exciting his moons would turn out to be. Planet-sized worlds in their own right, they are among the most fascinating objects in the entire solar system. We found out how exciting they are when *Voyager I* flew by in 1978.

As the closest large moon to Jupiter, Io is so stressed by Jupiter's gravity that its surface is constantly being "repaved" by volcanic eruptions; in fact *Voyager* actually photographed a sulfuric eruption in progress. Further from Jupiter is Europa, whose surface, while also relatively fresh, is solid ice. In 1997 the *Galileo* spacecraft sent back pictures suggesting that the ice ranges from 12 miles to less than a mile in thickness, and that beneath it might lie a planetwide ocean. A central task of future space exploration might be to search for simple life forms thriving within that ocean. The outer large moons, Ganymede and Callisto, have ancient surfaces scarred with impact craters. Jupiter has, as of the summer of 2002, 39 discovered moons.

Saturn: Lord of the Rings

No planet looks more stunning through a telescope than Saturn. Its magnificent rings are breathtaking, and have been since Christiaan Huygens discovered them in 1656. He announced his discovery as an anagram that deciphered to this Latin sentence: *"Annulo cingitur, tennui, plano, nusquam coherente, ad eclipticam inclinato."* (Translation: It is surrounded by a thin, flat, ring, nowhere touching, inclined to the ecliptic.) In 1979, the *Voyager* spacecraft studied these rings closely, recording hundreds of tiny ringlets grouped into six major ones. They are the result of at least one comet, asteroid, or moon that got too close to the planet and broke apart. In the 19th century Edouard Roche, a French mathematician, suggested that if an object that was loosely held together approached to a certain distance from a planet, it would be torn apart by the planet's tidal force. Sometime in the past, these objects passed within Saturn's "Roche limit" and broke apart to form the system of rings.

Saturn's Moons

In addition to its rings, Saturn has at least 22 known moons. Titan, the largest, is more than three thousand miles across and is the only satellite in the solar system known to have an atmosphere. Dione and Rhea resemble our moon; Mimas has a huge impact crater named Herschel, the result of a collision that almost destroyed it; Tethys displays a similar large crater. Phoebe orbits in retrograde motion, a motion opposite to that of the planet's rotation. It is rare for a moon to orbit this way around its master world; Phoebe probably does so because it was not formed along with the rest of Saturn's system of moons; it may have been a distant asteroid, or even a large comet, that wandered past until Saturn's gravity captured it.

On January 22, 1967 *The New York Times* published this splendid poem about Phoebe:

> Phoebe, Phoebe, whirling high
> In our neatly plotted sky,
> Phoebe, listen to my lay,
> Won't you whirl the other way?
> Never mind what God has said,
> We have made a law instead.
> Tells each moon where to go each night,
> Phoebe, won't you get it right?
> I'm afraid, little moon, you'll have to change.
> For Really, we can't rearrange
> All our charts from Mars to Hebe
> Just to fit a chit like Phoebe.

Uranus: Green Giant
The world that Herschel discovered in 1781 is now called Uranus, a giant, green planet, unique in that it rotates on its side! It is tilted at an angle of 98 degrees to the plane of its orbit. As a result, the planet has some strange characteristics. During much of its 84-year orbit around the Sun, either one or the other of its poles in turn faces the Sun and becomes the warmest place on the whole planet for up to 42 years.

Uranus has a dense atmosphere similar to that of Jupiter. Its rotation period of about 17 hours was first suggested by Steve O'Meara, an amateur astronomer of Cambridge, Massachusetts, based on his own visual observations, and was not taken too seriously until it was confirmed by observations of the spacecraft *Voyager 2*. The planet's temperature may be as low as −330 degrees F, a temperature at which its supply of ammonia would exist in the form of ice crystals. In 1977, an occultation by Uranus of a star led to the discovery of 9 rings, a number that increased when *Voyager* visited. Of its retinue of 21 moons, Ariel has some huge scarps that might be the remains of some violent past activity. Miranda is even more unusual, with fracture patterns and sudden landscape changes. Could that moon have fallen apart and then reassembled after some collision in its early history?

Neptune: Blue Giant World
Viewed from deep space, only two of the solar system's nine planets appear blue, the Earth and Neptune. Earth is blue from its abundance of water; Neptune, with four times Earth's diameter, is blue from its abundance of frozen particles of ammonia. Even though the planet is almost twice as far from the Sun as Uranus, it is not colder; possibly, Neptune is radiating heat from its core. The Neptune system orbits the Sun in about 165 years. In the year 2011, the planet will finally have completed just one orbit from the day it was discovered!

Of the planet's eight known moons, Triton is the largest. A frozen world with no atmosphere, Triton travels around Neptune in a retrograde orbit. *Voyager 2* discovered a 250-mile-wide moon, now called Naiad, which had been missed before because it orbits so close to the planet.

Pluto: Dark World at the Solar System's Exit
Distant Pluto, unknown to us until Clyde Tombaugh discovered it in 1930, is one of the strangest worlds of the solar system. It is 1400 miles in diameter, not nearly as wide as the continental United States. Its orbit is much more elliptical than any of the other planets, so much so that at its 1989 perihelion, or closest point to the Sun, it was closer to the Sun than Neptune! Its one moon, Charon, is so close by that it orbits Pluto in only six days.

PLANETS AND MOONS

Pluto orbits the Sun in the midst of a huge family of comets commonly called the Kuiper belt, after Gerard Peter Kuiper. (Frederick Leonard in 1930, and later Kenneth Edgeworth and Fred Whipple, also theorized this belt, which surrounds the solar system as a form of inner boundary.) An outer boundary, called the Oort cloud after Jan Oort, surrounds the solar system at a much greater distance. Does Pluto hold the key to the secrets of this little known segment of deep space? Of all the planets, it is the only one not yet visited by a spacecraft. Perhaps in the new millennium, a spacecraft will head out there, and provide us with answers.

Hints on Observing the Planets

The most interesting part of Earth's Moon to observe is its terminator, the twilight zone between its bright and dark portions. It is there except at full and new moon. It is the region of changing shadows, where craters and mountain ranges stand out in stark relief. Around full moon is the time to study the rays of big craters like Copernicus and Tycho. Stretching long distances, the rays are made of material blown out from impacts, and they testify to the power that comets and asteroids have to shape the worlds they hit.

Venus

Through a small telescope Venus is blindingly bright, and, like the Moon and Mercury, it has phases. Venus' thick covering of clouds prevents us from seeing its surface, and so besides the phase, a telescope view usually shows no detail. When Venus is in its crescent phase, you may try to see a faint glow on the darkened part. Known as the Ashen light, this glow is difficult to see.

Mars

With elegant names for its surface features like Elysium, Hellas, and Tharsis, Mars is an elegant planet to observe.

The Syrtis Major, for example, is a feature that looks somewhat like Earth's India, and the dark circular feature Solis Lacus is more popularly known as the "eye of Mars." But all these features do not face us together any night of the year, not even when Mars is physically above the horizon. We see the red planet in an array of sizes that depends on its distance from us. If the planet is near its aphelion, of farthest point from the Sun, it remains small enough that most observers will see very little with small telescopes. When the planet is close to us, we can see its changing array of features as it rotates once in about 24.6 hours.

Jupiter
The easiest observation to make with Jupiter is the changing positions of its four Galilean moons. Each night, the four moons are in different positions; Io and Europa orbit the fastest, and their motions are most readily apparent. The face of Jupiter is banded with dark lines called belts that are interspersed with bright zones. By far the most active belts are the two closest to the equator. The North belt is usually darker than the south one. Sometimes lanes of dark material called festoons jut out from a belt, or a dark bridge will temporarily join one dark belt with another.

Saturn
This planet's real highlight is the rings. You can also see Titan, the solar system's largest moon, and perhaps a few other moons. Unlike the moons of Jupiter, which seem always in a line, Saturn's moons can be above or below the planet. As Saturn orbits the Sun in a little more than 29 years, the angle at which the rings are presented changes slowly. The rings can be spread out to edge-on and invisible. The cycle takes place twice during each of Saturn's 29-year orbits.

PLANETS AND MOONS

Uranus, Neptune, and Pluto

It is not normally possible to see detail on the three outermost worlds, although Steve O'Meara succeeded in measuring Uranus's rotation period. Using the 9-inch refractor atop Harvard College Observatory, he waited several years before finally, on a twilit evening in 1984, he noticed two bright spots. He monitored the spots as they moved around the planet's poles (Uranus is tilted), and derived a rotation period of 16.2 hours. Two years later *Voyager 2* observations confirmed that value.

With Neptune and Pluto, the main thing is to identify them. Neptune is an eighth magnitude bluish star, but Pluto is about 6 magnitudes fainter. The best way to identify Pluto is to observe it over several nights, recording its changing position among its field's background stars.

REFERENCES

Copernicus, Nicholas, *De Revolutionibus Orbium Coelestium*, 1543; rpt. Ex Officina Henricpetrina, 1566.

The New York Times [New York] 22 Jan. 1967.

Ryan, James, *The New American Grammar of the Elements of Astronomy*. (New York: Collins and Hannay, 1831, 320.

Alfred, Lord Tennyson, "In Memoriam" (XXI, lines 17-20) *Victorian Poetry and Poetics*, ed. Walter E. Houghton and G. Robert Stange (Boston: Houghton Mifflin Co., 1968) 51.

CHAPTER 4

Planets of Distant Suns

"First to report discovery, Cole of Spyglass Mountain famous in a night."
—Arthur Preston Hankins,
Cole of Spyglass Mountain

One summer evening over dinner in 1960, Dad shared with us a story he had read when *he* was a youngster. It was *Cole of Spyglass Mountain*, Arthur Preston Hankins' novel about a boy whose love of the sky led him to observe Mars through his small homemade telescope. The novel ended with Cole finding evidence of life on Mars one night and becoming an instant celebrity. "Now that you're interested in astronomy," Dad said, "if you ever find that book in your wanderings through astronomy libraries, I'd love to read it again."

Five years later I began my long search for comets. As the years went by and Dad grew older, he would often ask if I had found *Cole of Spyglass Mountain*. Then with the onset of Alzheimer's disease, his memory began to fail. As his time grew short, he forgot names, people, and once even forgot that I was his son. But somehow he never forgot that euphoric story about a boy and a mountaintop that had captured his imagination so many years earlier. When I finally found my first comet in 1984, I felt as though I had rewritten *Cole of Spyglass Mountain* in a private way just for Dad. But he was too ill to appreciate it, and he died only a few months later. I will always regret that I could not share the moment of my first comet with him.

Life on other worlds was a subject Dad was always excited about. I think he would have been interested to learn of the two planets that have been found circling the star 47 Ursae Majoris, some 42 light years away. That sun, just below the bowl of the Big Dipper, is faintly visible to the naked eye. When I was a teenager I was assigned a small area of the sky to search each night with binoculars for possible comets. I have been familiar with 47 Ursae Majoris since I first looked at it on May 1, 1964. Now I learn that at least two planets, one at least twice as large as our own Jupiter, orbit that star. Is it possible that someone on that planet has been looking back?

Early Extrasolar Planet Hunting
In the early 1960s, some astronomers believed that the reason Barnard's star was showing a slow wobble was that a planet ten times the size of Jupiter might be affecting its motion through the sky. That conclusion probably is not correct, and the field of extrasolar planets was not seriously pursued until late in 1983, when the international Infrared Astronomical Satellite (called IRAS) was launched from California's Vandenberg Air Force Base. I saw the plume of flame from the roof of the Flandrau Planetarium in Tucson, hundreds of miles away, as the rocket carrying the new satellite roared into the night. IRAS mapped the entire sky in the infrared wavelength designed to detect sources of heat.

Since it was beyond our own atmosphere, IRAS was able to image sources of heat that could not be observed from the ground or from other means. One object it recorded was an extended area around the star Beta Pictoris. In 1984, the University of Arizona's Brad Smith used a coronagraph to block out the bright light from the star Beta Pictoris, enabling his telescope to record a large area of delicate nebulosity surrounding the young star. This cloud could be the early stages of the formation of a new system of planets.

Pulsar Planets
The best place to look for planets, one would think, would be around stars like the Sun. But that's not where the first strong evidence of Earth-type planets came from. Instead, these worlds orbit the burnt out remains of a star that had probably gone supernova many thousands of years ago. All that is left of this once blazing sun is a small neutron star that pulses as it spins very rapidly. This "pulsar" goes only by the name of B1257+12, a code representing its position in the sky as Right Ascension 12h 57m and Dec +12 degrees. In 1991, Alex Wolszczan and Dale Frail found evidence of three planets, each one about the size of the Earth, orbiting about the spinning remains of this star (Croswell 140-45).

These mysterious worlds pose some important questions. Assuming they existed when the star was a normal sun, what were they like? Is it at all feasible that life might have existed there? And what would it have been like to be on that planet, witnessing the instantaneous collapse and explosion of its life-giving star into a supernova? In any event, the blast would have instantly snuffed out any life forms on those worlds.

Tug of War in Space
Neither Geoffrey Marcy nor Paul Butler will forget that day in 1995 when they made their first announcement of a world

circling another sun. "We rarely knew which TV network we were being whisked to, from hour to hour," Marcy remembers. "We just smiled into the cameras and described the wonderful planets."

Ten years earlier, Marcy and Butler had joined forces to begin a search for planets that orbit other suns. Figuring out how they were to detect the presence of these worlds was a daunting task. These planets are too faint to be viewed directly with present technology, but maybe the way they affect the suns they orbit around could be detected. It was a new approach that makes use of Newton's old law: "For every action," his third law says, "there is an equal and opposite reaction." As the Earth travels around the Sun, the Sun swings very slightly in the opposite direction. It's like a child's game of tug of war in which two sisters grab opposite ends of a rope. As the younger, smaller sister pulls hard on one end of the rope, the older sister bends backwards only slightly to keep her balance. Our Sun plays a similar tug of war game with each of the nine planets, including Earth. Most of the Sun's wobble is caused by Jupiter, by far the most massive of our solar system's planets. When we look out into space, we cannot directly see distant worlds playing tug of war with their suns. We can tell if the game is being played, though, if the star is wobbling.

How did Marcy and Butler plan to eavesdrop on these distant games? A star's wobble can be detected when astronomers measure the speed with which a star is moving toward us or away from us. Their idea: Measure the spectrum of the star around which a planet might exist. If the spectrum is shifted to the blue, then the star is moving toward us; if to the red, then it's moving away from us. To make the calculation, light from the star must pass through a chemical vapor that will absorb specific wavelengths of light. These "spectra" would allow the detection of changes in velocity as small as the speed of a runner, a few miles per hour. Paul Butler's first task was to find a chemical that could produce results of that accuracy. One possible candidate: thiophosgene, which happens to be closely related to phosgene, the mustard gas that was used in World War I. Before he was faced with using such a dangerous chemical, however, he found that iodine worked just about as well, and it is so much safer. With the spectrum of this chemical as a calibration, in 1987 the team's search for planets began in a wave of uncertainty. This was completely new science, and the team was not certain of detecting the tiny wobble of a star that gives away the presence of an orbiting planet.

51 Pegasi: A Star like the Sun

The search for planets around other suns next made the news in 1995, when Michel Mayor and Didier Queloz of the Geneva Observatory, discovered a planet circling 51 Pegasi, a nearby star. Unlike the faint pulsar, 51 Pegasi is a bright star easily visible with binoculars, in the constellation of Pegasus. The planet's orbit is strange, lasting only four days and five hours! For the planet to circle the star that fast—the closest planet to the Sun in our solar system is Mercury, which takes 88 days to complete one revolution—that world would have to orbit in the outer atmosphere of the star.

In 2001, Marcy and Butler, now working with Debra Fischer, announced the discovery of a new solar system around 47 Ursae Majoris, more like our own than any other seen to date, a system of two planets that orbit the star in almost circular paths. Why is this important? From 1930, when Pluto was discovered, to 1992, we knew of no new major planets—anywhere. Since then our known roster of planets, and whole solar systems, has grown exponentially. As of the late summer of 2002, 88 stars had known systems of planets, and 11 of these had more than one planet. The total number of extrasolar planets was 101, and rising quickly. However, most of these planets circle their suns in wildly elliptical paths, rushing out to great distances, then closing in on them. If our solar system had planets like these, our hapless Earth would have long ago been destroyed in a collision with one of them. In this newly discovered solar system, the worlds orbit in almost circular orbits much like Jupiter and Saturn, leaving plenty of room for safely orbiting smaller worlds that could have life.

47 Ursae Majoris is the type of peaceful system astronomers have been looking for. Lately, they have been finding worlds both large and bizarre: One has a planet 17 times the mass of Jupiter, peacefully lumbering around a star in the constellation of Serpens. Off in another corner of the galaxy are two planets doing a cosmic dance around a small red sun just 90 trillion miles, or 15 light years, away. One orbits its sun in 30 days, the other taking exactly twice as long. They are in "resonance": Just like musical notes, these worlds have been forced by gravity into orbits that seem to harmonize with each other.

A Planet Transits a Star
Finding smaller planets, perhaps those as small as the Earth, might require a whole new approach. In 2000, David W. Latham of the Harvard-Smithsonian Center for Astrophysics analyzed the motion of HD (for Henry Draper Catalogue) 209458 and concluded that a companion, most likely a planet, orbited that star. The planet's orbit, however, happened to be oriented edge-on toward Earth. This meant that once each orbit the planet would pass in front of its star. During each pass, or transit, the light from the star should drop suddenly for a short while, and then rise again after the planet gets out of the way. It's like watching a distant eclipse; we are taking advantage of a geometric gift from nature.

In September 2000, the planet transited and gave itself away as its star's light dropped exactly the amount that Latham had predicted. This convenient new method allows astronomers to learn far more about the nature of the new planet than would be revealed by the other methods. For example, from the transit, the astronomers calculated that the planet circling HD 209458 is 1.27 times the size of Jupiter. We even know that the planet's density is less than water, so that this world, like Saturn, would float in an even larger body of water.

Observing planets transiting their stars could offer a gold mine of information about these new worlds. However, only those worlds oriented toward Earth will work; if the orientation isn't right, we never see a transit. Finding all these systems of worlds is vastly increasing our understanding of the nature of the stars around us. However, an ultimate and most satisfying goal is to find some day a world with intelligent life.

The Life Zone of a Star
Every star has a temperate region surrounding it in space. If a planet were to orbit within that region, its conditions might be favorable for the evolution of life as we understand it. We know of one successful planet in one solar system: Earth is fortunate to orbit within the Sun's life zone. In chapter three we saw how Venus and Mars, which orbit outside of the Sun's life zone, have atmospheric conditions that are hostile to life.

Might there be other planets, in our galaxy or beyond, that harbor life? To answer this question, we need to consider that only a small fraction of other stars are the size and type of the Sun. A brilliant blue supergiant star like Rigel, in Orion, might have a life zone, but a planet would have to be many times the distance between Earth and Sun to orbit within it. Rigel's strong ultraviolet radiation would give additional problems for life. Even our own Sun has UV radiation, but thanks to our protective ozone layer in Earth's atmosphere, life can thrive here despite the harmful UV rays. It is also appropriate to point out that our Sun is not the ideal star to support life. A slightly redder Sun would emit fewer UV rays, thus allowing life to evolve without the need for UV protection.

The Drake Equation

If we believe the message of science-fiction TV shows, intelligent, English-speaking life flourishes on many planets. The reality is that no intelligent life forms have yet been discovered on any other worlds. Besides intelligence, life forms would need to have some way of communicating with other planets to be detected. We have no way of knowing, for example, what humpback whales tell each other. It is possible that the galaxy is teeming with other intelligent species; it is even possible that they know about us. It wouldn't be hard: We've been freely advertising our place on the Sun's third planet since the Italian physicist Guglielmo Marconi sent a wireless message consisting of long-wave radio signals across the Atlantic in 1901. That simple message traveled across an ocean but did not stop at the other side; it went on around the world and into space. By the end of World War I it had reached 67 star systems, and by now it has passed through many hundreds more. (*The Royal Astronomical Society's Observer's Handbook*, pages 234-38 in the 2002 edition, contains a list of the nearest stars.) If any nearby stars have worlds with civilizations able to understand our every TV and radio transmission, they would learn about Hitler, our first flights into space, the Kennedy assassination, the Beatles, the Sydney and Salt Lake Olympics, and the events of a particular September 11. We have been advertising our existence to the galaxy for many years. Could anyone have been listening?

In 1961 the American astronomer Frank Drake considered that problem from a mathematical viewpoint that looked at a series of possibilities. The version I use here was adapted by Carl Sagan (247-48), and I admire it for its simplicity. However, I have added a new variable called f_j. The equation follows:

$$N = N_* \ f_p \ n_e \ f_j f_l \ f_i \ f_c \ f_L$$

We begin with N_*, which is the number of stars in our galaxy. Estimates range from 100 billion to 400 billion; for the purpose of this discussion we use 200 billion. All the rest are variables designed to focus our search.

The phrase f_p is the fraction of stars that have planets circling them. Every star might have some sort of planetary system. But say that only half the stars do, and then our search is now limited to 100 billion stars.

The function n_e represents "ecologically sound for life." It answers the question of how many planets orbit within the life zones of their stars. Although it is possible that avian life forms could exist in the atmosphere of a gas giant, we assume that these planets should not be like Jupiter, which might not even have a solid surface. If a tenth of the 100 billion suns with planets had at least one of this type of world, we'd be left with ten billion life-zone worlds.

f_j: Astrogeologist Gene Shoemaker once suggested to me that it is not enough for a planet to exist in a life zone. At least one large planet like Jupiter needs to exist somewhere in the system as well. In Earth's case, were it not for Jupiter being the solar system's vacuum cleaner, the Earth would still be a target for major impacts every few hundred years. No form of advanced life could evolve in such a case. So, in this book there appears a new constraint to an old equation, the fraction of solar systems with Jupiter-sized planets. If a tenth of the solar systems comply, we now have one billion possibilities. The Jupiter issue has a recently discovered side effect. It may be common that large planets, like the ones believed to circle 16 Cygni B and 70 Virginis, are in highly elliptical orbits. Such planets would probably wreak havoc on peaceful, Earth-type worlds trying to orbit in a star's life zone. In fact, were Jupiter not in a circular orbit, Earth and Mars would probably have been destroyed or flung out of the system long ago (Marcy and Butler, 1998).

The fraction f_l is the fraction of planets where life does take hold. The value of this fraction is the subject of hot debate: Some scientists think that life will probably arise on all these worlds; others say that the origin of life is a rare thing. Here we split the difference and suggest that life starts on half of these planets. We are left with 500 million worlds on which life has begun.

f_i: The difference between life of any sort and intelligent life is enormous. If we say that intelligent life evolves on only 1/100 of these planets, then f_i brings us down to five million intelligent worlds.

f_c: Our next consideration is one of communication. If a tenth of these intelligent worlds develop an ability to communicate with us through technology, then the number drops to 500,000.

f_L: Although we have the ability to communicate via radio and television, we've had it for only a century. This last part of the equation asks how long a technological civilization will last. In our experience, we developed the ability for technological communication just a few decades before we acquired the ability to destroy ourselves with nuclear arsenals. Is that a coincidence, or is the galaxy full of civilizations that have radio and that are on the brink of self-immolation? The fraction f_L is a measure of the political state of the galaxy. It would be wonderful if most civilizations do survive after reaching the capacity for self-destruction. Let us say, very conservatively we hope, that of all the civilizations we have postulated so far, a tenth of them have survived.

Thus, N = 50,000. We should be able to receive messages from no fewer than fifty thousand civilizations within our own galaxy.

Actually, all of the assumptions we have made could be hugely off. The galaxy could be teeming with millions of civilizations. It is far less likely that only one planet in our entire galaxy, Earth, has allowed life as we know it to evolve. A worldwide series of projects called SETI (Search for Extraterrestrial Intelligence) has been searching for extraterrestrial civilizations for some time. The effort uses a variety of radio telescopes to search for artificial signals from other worlds (Morrison, 1979). Although unusual signals have been picked up, none have been confirmed, so far.

Ockham's Razor and UFOs
Are there such things as flying saucers from other worlds? If all the reports of such things are real, then Earth has been explored, probed, studied, and dissected every day by many alien races. While such reconnaissance is certainly possible, almost every scientist doubts that UFO reports are the results of alien visitors.

Science works with several basic principles, one of which is Ockham's Razor, also known as the law of parsimony (the adoption of the simplest assumption in forming a theory). It was first developed by the fourteenth-century Franciscan William of Ockham. He suggested that where two or more explanations fit the observed facts, the simplest one is probably the correct one: "What can be done with fewer is done in vain with more." We can apply the old Oxford professor's wisdom to UFOs. For any unconfirmed sighting of something in the sky, the most complex explanation would be that light is the product of an advanced civilization that took the trouble to build a spacecraft and use it to travel all the way for years, just to visit us. We must first seek a simpler explanation. It has been my experience that Venus, when it is bright in the evening sky, has often been accused of being a UFO, as have planes landing at an airport. I used to live in Corona de Tucson, a town situated at a 180 degree angle to a distant airport runway. As planes approach at night with their landing lights on, they look like bright stars standing still in the sky. As the viewing geometry changes just before landing, they suddenly appear to pick up speed dramatically. One question you might ask the tellers of stories of rapidly moving lights in the sky is this: Was the rapid flight accompanied by a sonic boom? Anything flying nearby at faster than the speed of sound *must* leave such a noise; otherwise the object could not have been traveling faster than sound. These UFOs still have to follow the laws of physics.

Although explanations like these debunk many UFO sightings, there is a possibility that we have been visited in the past, or even are being reconnoitered right now. The galaxy is big, and our own analysis in this chapter suggests that it might have thousands of inhabited worlds. As we look at the sky on a dark night, we can only imagine what life might be like on these worlds. We can also wonder if we'll ever know the answer.

REFERENCES

Bishop, Roy. *Observer's Handbook*. Toronto: Royal Astronomical Society of Canada, 2002.

Croswell, Ken. *Planet Quest: The Epic Discovery of Alien Solar Systems*. New York: The Free Press, 1997.

Hankins, Arthur Preston. *Cole of Spyglass Mountain*. New York: Dodd, Mead, and Co., 1923.

Marcy, Geoffrey, and R. Paul Butler. "The Diversity of Planetary Systems." *Sky & Telescope* Mar. 1998: 36-37.

Morrison, Philip. *The Search for Extraterrestrial Intelligence*. NASA SP-419 ed. New York: Dover, 1979.

Sagan, Carl. *Cosmos*. New York: Ballantine, 1985.

CHAPTER 5

The Story of the Stars Above Us

> The kingly brilliance of Sirius pierced the eye with a steely glitter, the star called Capella was yellow, Aldebaran and Betelgueux shone with a fiery red. To persons standing alone on a hill during a clear midnight such as this, the roll of the world eastward is almost a palpable movement.
> —Thomas Hardy, *Far from the Madding Crowd*

Observing a meteor shower without a telescope is an ideal way of seeing a "palpable movement" of the stars as they cross the sky with the passage of a night. Watching the stars move across the sky as a result of the Earth's diurnal motion is the purest way of observing. You see the stars in much the same way the ancient Akkadians saw them more than five thousand years ago. These people lived on the land that eventually became Babylonia, and they, along with other peoples around the world, followed the daily motions of the stars around the pole, and the motions of the planets. All these peoples contributed to the great legends that line the figures we see in the sky.

One of the best known sets of legends deals with the seven stars we call the Big Dipper. While the ancient Arabs saw these stars as part of a Great Bear that circled the sky every night, old English legends see it as a plough. Some Native American tribes have the Never-Ending Bear Hunt, which includes the stars around the Dipper, such as the semicircle of stars called Corona Borealis, or the Northern Crown. The annual hunt begins in the spring, when Corona Borealis rises over the eastern horizon after dark. The semicircle is actually the mouth of the Bear's cave, and as he exits, seven warriors begin to hunt him down. Through spring and summer, the hunters ride through the sky, each star of the Dipper representing a hunter in search of the bear. As summer turns to fall, the hunters give up, and one by one they sink into the northwest. But just as the bear is about to set in the west, the last hunters finally catch up with him. With the arrival of spring a new bear exits the Corona Borealis cave, and the hunt begins all over again.

As seen from most of the United States, the Big Dipper never sets. As the Earth rotates about its axis, which points north and south, all stars, even the Sun, appear to circle the north or south poles. The stars farther from the poles rise and set, like the Sun. But depending on your latitude, some stars are so close to a pole that they neither rise nor set; they just appear to circle the pole star and are called circumpolar. At the equator, all stars rise and set; at the poles, all the visible stars are circumpolar. Among the constellations seen as circumpolar, Camelopardalis, the giraffe, seems most ill at ease, surrounded by two bears, a dragon, and a lynx!

Northern Hemisphere Sky in Spring

The Big Dipper is the key to getting familiar with the sky of spring evenings. Beginning with the two "pointer" stars at the end of the Dipper bowl, point northwards towards Polaris, the North Star. Join the three stars of the handle to form a curved line, or an arc, and "Arc to Arcturus," the bright orange star in the constellation of Boötes, the herdsman. From its distance of 34 light years, Arcturus was the star whose rays hit a photocell that turned on the lights for the 1933 Chicago World's Fair.

From Arcturus, "speed" to Spica, the bright blue star in the constellation of Virgo, the Virgin.

Summer Sky
As spring constellations gradually move toward the west, summer evenings bring new patterns with the rising of three bright stars. Vega is the brightest one. Its constellation is called Lyra, the Lyre or harp. Vega, a hot, blue star, is bright because it is only 26 light years away. Deneb is the brightest star in Cygnus, the Swan, a beautiful figure that straddles the Milky Way. Deneb is a giant sun that is intrinsically very bright but distant: it is some 1,500 light years away. To the southeast is Altair, the brightest star in Aquila, the eagle. Altair is 16 light years away. Vega, Deneb and Altair form the isosceles "Summer Triangle" with Altair at the apex.

Autumn
The autumn constellations of the Northern Hemisphere do not offer as many bright stars as the other seasons, but they do form a pattern that tells a story. Cassiopeia and Cepheus, the Queen and King of Ethiopia, had a daughter, Andromeda. Angered by Cassiopeia's incessant bragging about her daughter's beauty, Neptune captured Andromeda and chained her. At the moment she was to be devoured by Cetus, Perseus arrived to save her and carried her via Pegasus, the winged horse, into the sky. All the players in this story are represented by constellations in the autumn sky.

THE STORY OF THE STARS ABOVE US

Winter
Although *Orion*, the hunter, utterly dominates the sky at this time of year, winter is also the time of the year to enjoy the "Heavenly G." To find it, begin with Aldebaran, the bright red star in Taurus, and continue through Capella in Auriga, the charioteer, and the brightest stars of Gemini, Castor and Pollux. The path of the "G" continues through Procyon, the brightest star of Canis Minor, and Sirius, of Canis Major, the brightest star in the sky. The curve of the "G" turns in Orion at Rigel, and goes from Rigel to Betelgeuse.

Southern Constellations
Southern Hemisphere observers can see three magnificent constellations that are unavailable to northern viewers. Carina lies in one of the most beautiful regions of the Milky Way. It had a modest beginning as the keel of the ship that Argo built for Jason and his Argonauts as they sailed out to search for the golden fleece. The original Argo Navis was so vast, and covered so much of the southern sky, that astronomers have divided it into the three separate constellations—the Stern (Puppis), the Sail (Vela) and the Keel (Carina). Centaurus is Chiron the Centaur, whose brightest star, Alpha Centauri, is the closest star system to the Sun.

The most famous southern constellation is also the smallest in the sky. Crux, the Cross, was vital to sailors for centuries for the simple reason that it pointed the way to the south celestial pole, which has no bright star like Polaris to mark it. It is a part of the flags of several southern nations.

The Constellations in Modern Astronomy
With all their unusual shapes and histories, you might think that the constellations have no role at all in modern science. But in 1930, the International Astronomical Union decided that they do offer an easy way to divide the sky into specific

regions, and they approved a division of the sky into 88 constellations bordered by specific lines of *right ascension* and *declination*. Right ascension is the projection of Earth's longitude on the sky; declination is the projection of latitude. All novae, or exploding stars, within our galaxy are named according to the constellation in which they are found, as was Nova Cygni 1975, the most recent bright naked-eye nova.

Colors of Stars
Thomas Hardy's sky, as viewed at the start of this chapter, had stars of different colors. Colors and hues are important in astronomy, for we can learn a star's temperature and age from its color. Two of Hardy's stars, Aldebaran and Betelgeuse, are senior citizens, cooler and older than our Sun. How do we know this? Their secrets are given away when you notice that they're not white, but reddish. There are younger stars in the stellar family also; Vega, for example, is a bright bluish star almost overhead; blue stars are hotter and younger than the Sun.

A star's color is an indication of its temperature. Astronomers classify stars according to these rainbow colors, ranging from violet to red. After studying the spectral colors of more than 200,000 individual stars about a century ago, the Harvard astronomer Annie Jump Cannon divided them into different classes. Although she at first arranged them alphabetically, she later consolidated them and reordered them into seven basic groups covering stars from the hottest deep-blue to cool, red suns, and lettered O, B, A, F, G, K, and M. There are very few O stars; these deep-blue suns are so hot that they burn themselves out relatively quickly. The westernmost star in Orion's belt, Mintaka, is an example. B stars, like Rigel in Orion, are the hottest common blue stars. Vega, Deneb, Alpheratz in Andromeda, and Mizar, the middle star of the Big Dipper's handle, are bluish-white A stars. Polaris and the brightest star in Libra, called Zubenelgenubi, are F stars. Our Sun, 47 Ursae Majoris, and 51 Pegasi—all incidentally with planetary systems—are yellowish G stars. Arcturus is an orange K, and Betelgeuse and Propus (also called Eta Geminorum), belong in the red M class. The variable star TX Piscium is a representative of a very red "C," or carbon star.

The Hertzsprung-Russell Diagram

How do these classes of stars fit into a graph? One of the most important tools ever created in astronomy, the Hertzsprung-Russell (H-R) diagram, connects theory and observation to reveal the relation between the absolute magnitude of a star and its spectral type, or color. (Absolute magnitude is a way of measuring a star's actual luminosity, which does not depend on its distance from us.)

In 1911, the Dutch astronomer Ejnar Hertzsprung plotted the colors of a large series of stars in star clusters against their apparent magnitudes (brightness as we see them in the night sky) and then published a diagram showing the result. At the time, Hertzsprung's diagram was pretty much ignored. Two years later, Henry Norris Russell, a well-known American astronomer, produced a similar chart that included stars near the Sun, stars of which he knew the absolute magnitudes. Eventually the work became known as the Hertzsprung-Russell diagram. It is a graphical way of understanding the differences among types of stars, and how stars evolve from one type to another.

When the first diagrams were plotted, Hertzsprung and Russell independently found that almost 90 percent of the stars fell in a string ranging from hot blue stars at the upper left to cool, red stars at the lower right. This band of normal-range stars is called the Main Sequence. The brightest supergiant stars fell at the top, and a large number of hot, white dwarfs fell in the lower right end. Over the course of a star's lifetime, as it evolves it will appear first in the upper left of the diagram, and then "travel" from the upper right to the lower left.

From Birth to Death: Stars Throughout the Galaxy
"Bok globules" are a specific type of dark nebula that mark places where stars are in an advanced stage of formation. The Bok globules have a fascinating human history that goes back to the early 20th century, when the American observer Edward Emerson Barnard recorded them in his searches of the sky. They remained catalogued but not well studied until 1947, when Edith Reilly, a young technical assistant at Harvard Observatory and a victim of multiple sclerosis, asked to work on some project with the astronomer Bart Bok.

Although most of the assistants of that era had to handle large numbers of heavy photographic plates, Bok understood that Reilly did not have the strength to handle them. But he had an idea, and suggested that Reilly examine Barnard's catalogues of dark nebulae and prepare a list of candidates for further study. Reilly soon took particular notice of some small, round, and very dark clouds. Bok had noticed these types of nebulae before, but when Reilly called his attention to them he became intrigued, and decided to photograph some in greater detail. He was astonished at how dense these clouds really were. A zero magnitude star directly behind one of these clouds would be invisible even through a telescope as powerful as the Hubble Space Telescope. Bok believed that these objects were made of gases that were condensing under their own gravity in preparation for the start of fusion. As a search of the newly completed Palomar Sky Survey revealed 17,000 Bok globules, they came to be widely known among astronomers.

What would it be like inside a globule? A large amount of unlit gas—perhaps the equivalent of nine times the mass of the Sun—would pass through the globule during its 100-million-year lifetime, and some of this gas would be captured by the globule's own frigid particles. Over this time the whole globule would shrink. If the globule were small, it would form one or two stars; if it were larger it might form a cluster of stars.

In 1995, a team led by Jeffrey Hester of Arizona State University took a remarkable set of images near the center of the Eagle nebula, one of the Milky Way's great star-forming regions. In it they found a specific kind of globule that they called an evaporating gaseous globule, or EGG. It is a small, dark region filled with evaporating gas. Bok globules are typically less than a third of a light year in diameter, while EGGS are hardly more than a light week across. An EGG it seems, is the womb of a single star.

Protostars
Although it is very difficult to find a star in the process of forming, astronomers George Herbig and Guillermo Haro identified "Herbig-Haro objects," which are small nebulae that vary in brightness on an irregular basis (Reipurth, 1999). They are stars in the same state the Sun was in 4.5 billion years ago, when, just before it ignited, it began absorbing large amounts of the nebula around it and blowing away the rest. Once a star ignites, its radiation dissipates most of its nebula fairly quickly. A variable star called T Tauri, a hot blue star, is an example of a star that is in the early stages of stellar fusion.

Blue Supergiants
Mighty suns that are very luminous, the blue supergiants are among the brightest stars in the sky. The smallest blue supergiant is about ten times the Sun's diameter, but larger ones can grow to as much as a thousand times the diameter of our

Sun. Supergiants also burn up their hydrogen at a much faster rate than do stars like the Sun. One of the largest known supergiants belongs to a double-star system called Epsilon Aurigae, a star in the constellation of Auriga the charioteer. This stellar brontosaurus is almost 3,000 times larger in diameter than the Sun. If this star were in the Sun's place, it would envelop the orbits of all the planets as far out as Saturn!

Main Sequence Stars
The vast majority of stars belong to the Main Sequence, and they all owe their existence to the fact that they, like the Sun, convert hydrogen into helium. Stars on the Main Sequence stretch appear on the H-R diagram as a band. Our Sun is a Main Sequence star. It started out that way with only about 3/5 its present luminosity. As it continues to shine normally for another five billion years, it will gradually grow until it shines twice as brightly.

The Sun, as a Main Sequence star, works because it is in *hydrostatic equilibrium*. Its mass, forced toward the core by gravity, is precisely balanced by the outward pressure of nuclear fusion in its core. For a star to work, each layer must be in hydrostatic equilibrium, with gravity being balanced by gas pressure. In the star's outermost layers, where there is little weight pressing down, the gas pressure is low. Deeper in the star, weight and pressure increase. Energy is transferred by radiation, convection and conduction from the hot inner parts of the star to the outer regions, and then into space to light and heat planets like Earth. The more massive the star, the less time it spends in the Main Sequence. A star much more massive than the Sun might spend only a few million years as a Main Sequence star, but a low-mass red dwarf sun could last several hundred billion years. Since that is many times longer than the age of the Universe, red dwarfs essentially last forever. They are the most common kind of star, but because they are so small and faint, we see only those that are closest to us. Proxima Centauri, the part of the Alpha Centauri system closest to us, is a red dwarf.

Red Giants

Each tiny section of a relatively cool star like Antares radiates very little energy. However, because of its large area, Antares ends up being very luminous. Even though red giant stars have much the same surface temperature that the Sun has, they are some ten times larger in size. Hind's Crimson star, also called R Leporis, is an extreme example. A carbon star south of Orion in the constellation of Lepus, it is so red that some observers liken it to a drop of blood in the sky. We will learn more about red giants later, when we explore the future of the Sun.

White Dwarfs

A different example of stellar behavior from red giant, a white dwarf star is only the size of the Earth. These stars are very hot, but because they are so small they are quite faint. They represent a late stage in the evolution of a star in the Sun's size range. By the time they have reached the white dwarf stage they have lost at least half their original masses. The matter inside a white dwarf is not really gas anymore, at least not gas as we understand it in daily life, but a different form called *degenerate matter*. The nuclei of atoms in such a star are kept apart by a swarm of electrons that fight against the incredible gravity that is collapsing the star.

The most famous example of a white dwarf is the companion of Sirius, the brightest star in the heavens. Sirius is often called the Dog Star because of its prominent location in Canis Major. Less than 9 light years from Earth (or 51 trillion miles), it has 28 times the luminosity of the Sun. In 1844, Friedrich Wilhelm Bessel discovered that Sirius was displaying alterations in its motion through the sky that indicated that something was pulling on it, perhaps a companion with an orbital period of about 50 years. In 1862, Alvan Clark, the century's most famous maker of telescopes, was testing a new telescope objective lens on Sirius when he detected a faint companion almost lost in the glare of the brilliant main star. Nicknamed "the Pup," this interesting star has fully a third the mass of its famous companion and 85 percent of the mass of the Sun. However, it has only 1/360th of the Sun's luminosity and—amazingly—is only the size of the Earth! The Pup is an extremely dense star—more than two thousand times the density of platinum—in which a teaspoonful of material weighs a ton! We call it a degenerative white dwarf; its degenerate matter is as hard as steel and subject to very high pressure.

Is it possible that several billion years ago Sirius's companion was a normal star like the Sun? Its mass suggests that it was, and that from its perch hidden in the glare of the sky's brightest star, it is telling us about what our Sun might be like 10 billion years in the future.

Double Stars

On any Northern Hemisphere clear night, but especially in late winter, spring, and summer, you should easily be able to spot the Big Dipper. Look carefully at the middle star of the handle: nearby is a faint star. For many beginning observers of the night sky, Mizar and its faint "companion" Alcor are the first example of a double star in space. Mizar and Alcor, however, do not form a true double system. Mizar is 59 light years away, Alcor a much more distant 82. Such a pair of stars is an *optical double*, meaning that the two stars involved only appear to be near each other because they are on our line of sight. However, Mizar itself is a true double system, its two components revolving around each other. Through a small telescope, Mizar divides into two separate stars.

THE STORY OF THE STARS ABOVE US

Orbits of Multiple-Star Systems

One clear evening, an enthusiastic new observer was shown a double star through the wonderful 6-inch refractor of the Royal Astronomical Society of Canada's Montreal center. When the operator of the telescope explained that the two stars actually orbit one another, the observer answered with an excited "Yes! I can see them moving!" (Williamson, 1962) This is unlikely. The two stars involved have an orbital period of not less than 600 years. Relative to the distances from the Sun, most double systems are at great distances from each other. Also, since the masses of the two components are often comparable, the two stars clearly revolve around a common center of gravity rather than one revolving around the other. Sirius and its companion revolve around each other in almost exactly 50 years, as we have seen. The two brightest components of Castor, the brightest star in Gemini, the twins, take 380 years to complete a revolution.

Alpha Centauri, the nearest star to the Sun, is a triple system. The faintest one (and the one closest to us), Proxima Centauri, is a red dwarf; the two other, bright members orbit each other once every 80 years. The orbits of multiple systems can get very complicated. Zeta Cancri is an interesting example. Its two brightest members are each double systems; one has a period of 60 years, the other 18 years. However, the two pairs mutually revolve about each other in a period of more than 1,100 years. Castor is another case; it is actually a sextuple system. In addition to the two main stars orbiting each other, each main star has a small companion, and two other faint stars orbit each other nearby. One of the prettiest multiple systems I know of goes by the name of Sigma 2816 Cephei. Nearby is Sigma 2819, a binary system. They are part of an open cluster called IC 1396 in Cepheus.

In chapter 6 we will learn of doubles with much shorter periods of variation. These stars are so close to one another that we learn of their nature when one member passes in front of the other. Algol in Perseus is a double-star system in which the fainter star eclipses the brighter one every 2.4 days.

The Life of the Sun
In chapter one, we went back in time and explored the primordial solar nebula. As it grew smaller, particles colliding with one another heated up its center. The temperature of the center slowly rose as the molecules collided more often and more violently. When the temperature was high enough for hydrogen to fuse into helium, the Sun ignited and began its life as a star.

What happened next? At that moment the Sun brightened enormously, and the immense weight of its outer layers trying to collapse was precisely balanced by the pressures created by the nuclear fusion reactions taking place in its core. The Sun has maintained this state of hydrostatic equilibrium for some five billion years, pouring four million tons of hydrogen every second into its process of making helium.

The Sun's life will get complicated when its core exhausts its supply of hydrogen, an event that should take place some five billion years into the future. The events that follow happen relatively quickly, within a few million years. First, the hydrogen in the Sun's outer layers, or shell, will begin to fuse. Like a rapidly spreading fire, the fusion process will then move progressively through its outer layer, causing them to rise in temperature and expand. Over several hundred million years, the Sun will swell in size until it becomes a red giant.

At some point the Sun's fusion process will reach its cooler outer layers, where the temperature is about ten million degrees. Below that temperature, fusion does not take place, so the fusion process in the Sun's core, though not in its shell, will stop. The hydrostatic equilibrium will end, and for the first time in ten billion years, the Sun will start to collapse. The helium atoms in the core will quickly become very compact, raising the Sun's temperatures so much so that the helium begins its own nuclear fusion. Helium fusion can begin gradually, or suddenly in an explosion called the helium flash. The time from the onset of helium fusion to the core explosion could be but a few hours. Because the helium flash involves an explosion deep in the core, almost none of the energy of the explosion reaches the outermost layers of the Sun. In fact, after the helium flash, the Sun's temperature drops, and the fusion of hydrogen in the Sun's shell slows down dramatically. The Sun begins to contract. For the relatively short period of ten thousand years, the Sun shrinks in size and rejoins the Main Sequence. By this time, hydrostatic equilibrium is restored once more. The helium fusion process now produces carbon and oxygen at the core. Eventually, helium starts to burn in the atmosphere of the Sun. Once again, the Sun swells into a red giant, but this time it takes only a few million years to do so. As the Sun turns red, it will expand so much that it will envelop Mercury, Venus, and possibly even the Earth.

During the Sun's red giant phase, it might develop a layer in its outer atmosphere that absorbs heat energy coming from its core. If that layer evolves, then the Sun will become a variable star, each cycle lasting from several days to several months. As the Sun contracts in size it will brighten. At a certain point, it will then stop contracting and expand in size once more. During the expansion, it will fade. It is hard to predict how drastic the variation will be. In the constellation of Cetus is a star called Mira, the wonderful. If Mira's behavior is any indication of what is in store for the Sun, our star could change its light output by a hundred times every several months.

After a further passage of millions of years of time, all of the Sun's helium will have fused to carbon and oxygen. With nuclear reactions stopped again, hydrostatic equilibrium is lost and the Sun will resume its slow collapse. In this last attempt to continue life as a star, the Sun will pulsate, and throw its outer layers into space.

With the Sun's outer layers spread throughout what's left of the solar system, all that remains is its core. Over many billions of years, the Sun will then radiate its heat away, and it will grow cold and dark, becoming a black dwarf. This event is very far off in the future; in fact, our galaxy may not be old enough yet to have any black dwarfs. During this long, slow process, the Sun's remaining planets will still circle it. No longer viable and beautiful, the planets, like their Sun, will be lifeless and bleak, their story over.

REFERENCES

A General Catalogue of Herbig-Haro Objects. 2nd ed. Ed. Bo Reipurth. 1999. 22 Aug. 2002 <http://casa.colorado.edu/hhcat/>.

Hardy, Thomas. *Far from the Madding Crowd*. London, 1874.

Williamson, Isabel K. Personal Interview. 1962.

CHAPTER 6

Variable Stars

Wake! For the Sun, who scatter'd into flight
The Stars before him from the Field of Night,
Drives Night along with them from Heav'n, and strikes
The Sultán's Turret with a Shaft of Light.

Before the phantom of False morning died,
Methought a Voice within the Tavern cried,
"When all the Temple is prepared within,
Why nods the drowsy Worshipper outside?"
—Edward FitzGerald,
The Rubáiyát of Omar Khayyám

A temple indeed, the night sky is peppered with miracles of every description. Some of the best are the stars that change in brightness, performing before your eyes! Some of them change because they are really double systems in which a faint star passes in front of a bright one, causing the total system brightness to drop. Others vary because of some change in the astrophysical nature of the star; as they expand and contract, their magnitudes fall and rise. Some stars vary explosively, rising many times their original brightness within just a few hours.

I became interested in variable stars some forty years ago, and although I always knew that variables were among the sky's richest treasures, my early meetings with some variables were frustrating experiences. I had no idea how bright Chi Cygni would be, for example, when I began my search for it. At its best, this slowly varying red giant is faintly visible to the naked eye in Cygnus, the swan. But the night I met Chi Cygni, that ancient sun was not at its best. For two hours I worked with star charts until I finally located its dim red glow. Almost at the limit of my 8-inch telescope, the star was about 200 times fainter than what I had hoped for. The good part was that the star had nowhere to go but up, and over the next few months it slowly increased in radiance till it was easily visible through binoculars as a bright reddish beacon.

A second star in Cygnus, which I met two years later, was a lot easier. On the last Saturday night in August, 1975, I observed what I thought was a satellite, but I quickly noted that it did not move. There, shining brightly in the center of the Milky Way, was a blazing new star! Later called V1500 Cygni, this star was undergoing an explosion that blew off an incredible amount of hydrogen it had patiently collected, over tens of thousands of years, from a silent and invisible partner. Normally buried in the distant depths of the Milky Way, V1500 Cygni was a faint star visible only in the world's most powerful telescope. But on that summer weekend it soared above the others, becoming for a brief time one of the brighter stars in the sky. To be able to catch its act was truly a highlight of my observing life.

Observing Variables
Ptolemy was wrong in more than one way when he invoked his geocentric theory of the Universe, which held the last, distant region of sky as a sphere of fixed and unchanging stars. Although the distant stars may appear not to move, they all do; it just takes many tens of thousands of years to

see an appreciable change in the sky. However, it takes only a week or so to see that some of these distant stars change their light output, in some cases dramatically.

The fourth brightest star in the constellation of Cepheus is one of the most famous examples. Delta Cephei is an intrinsically luminous star that goes through a cycle of variation of its light every five days. As its outer shell expands in size, the star decreases its energy output and fades in brightness, but only to a certain point. The star then begins to contract, and its brightness rises relatively quickly. The whole cycle finishes in precisely 5.37 days. This star may have contributed to the death of John Goodricke, a deaf-mute astronomer from London, in 1784. His career soared after his discovery that Algol, an eclipsing binary star in Perseus, varies in just over two days. He went on to discover the variation of another eclipsing binary called Beta Lyrae, a star not far from Vega. But in 1786, after several weeks studying the regular variations of Delta Cephei under damp English nights, he developed pneumonia and died at age 21. Goodricke's immense contribution was to show how disabilities need not prevent great discoveries; the secrets of some of the most famous variables in the sky were first revealed to his keen eye and sharp mind.

The Magnitude Scale
Stars appear in all different brightnesses, which we call magnitudes. The first known magnitude scale was invented possibly by ancient Jewish philosophers, who determined that the Sabbath and other religious festivals should not be considered over until at least three stars appeared in the sky. Their aim was to ensure that a festival was not ended prematurely if the Sun, though still above the horizon, had set behind a mountain or other obstruction. Their scale had three divisions; the brightest stars, some of which could conceivably be seen in daylight, did not qualify, nor did the

faintest ones, which appeared long after sunset. The sighting of three stars in the middle brightness division would be the proper signal that a festival had ended.

Our modern system of magnitudes dates back to the second century B.C. Greek astronomer Hipparchus, who divided the stars into six brightness groups, with the 20 brightest stars designated as first magnitude, and the faintest stars as sixth magnitude; the higher the magnitude number, the fainter the star. The 19th-century astronomer Norman Pogson refined the magnitude scale by defining Polaris, the north star, as being second magnitude. He assigned a difference of 100 times between a zero magnitude star and a fifth magnitude star; thus, each magnitude is 2.5 times fainter than the next lower number.

Many stars change their light output, and while it might be fun to read about them, there's no substitute for seeing them perform. The Cepheids are just one type. Algol is the best known example of the type we call an eclipsing binary. It stays at constant brightness for two days, until its faint companion passes directly between it and Earth. Over the next few hours, the star fades to less than half its original brightness, stays there briefly, and then, as the companion moves out of the way, the system gradually recovers its full brightness. Isn't it amazing that from your own back yard, with no optical aid whatsoever, you can look up and watch the orbital mechanics of a star 815 light years away?

Cepheid variables change brightness as they expand and contract on a regular basis; as they expand they fade, and as they contract they brighten. And therein lies a story. Imagine a small room with a wood desk, a set of photographic plates, and a woman in a turn-of-the-20th-century long dress peering into a magnifier. The woman's name was Henrietta Leavitt, and she was studying the brightness changes of distant Cepheid variables. It is hard to conceive of a routine assignment leading to a greater breakthrough.

VARIABLE STARS

Leavitt built upon the work of a colleague, Solon Bailey, who had noted large numbers of Cepheid variable stars in globular clusters. She studied some 25 Cepheids in the Small Magellanic Cloud, one of the closest galaxies to us, and found that the brighter their average magnitudes were, the longer were their periods of variation. Since all the stars in the distant Magellanic Cloud are about the same distance from us, the astronomer Harlow Shapley later concluded that the stars display a correlation between their periods of variation and their average magnitudes. He turned this relationship into an astronomical yardstick for measuring distances.

The distance yardstick of the Cepheid stars is now being extended farther out into space. With the Hubble Space Telescope, astronomers are able to peer out and spot Cepheids in more remote galaxies.

The behavior of the red giant stars is more complex, varying over months, not hours or days. Mira, also known as "the wonderful," is the most famous example. This star reminds us of what our own Sun might become many billions of years from now. Although Mira might have been a star like the Sun once, it is no longer. The change is tremendous: over eleven months, Mira fades from being an easy-to-spot naked-eye star to a red speck barely visible through a small telescope. Moreover, the star's behavior cannot be accurately predicted. Its brightness may fall faster on one cycle than it did on the previous one, or it may take its time rising from its minimum brightness. I have watched R Leonis, another Mira-type star, shine at a constant magnitude for weeks at a time before resuming its variation.

The most exciting variable stars are the irregular ones, because their behavior cannot be predicted. My favorite example is TV Corvi, a star in Corvus, the crow. I first learned about this star while conducting research for a biography about Clyde Tombaugh, the discoverer of Pluto.

While checking the voluminous notes he wrote on the envelope for each photographic plate he took during his search for planets, I read of his discovery of a nova on a plate where no star was seen on the two earlier plates. Although Clyde reported the star's outburst to his superior at Lowell Observatory, it was never followed up. I decided to investigate: At the Harvard College Observatory is one of the world's best archives of photographic sky surveys. But although they had a plate taken of the Corvus region a few days after Clyde's exposure, the nova did not appear. Discouraged, I looked at some other pictures of Corvus taken over many years. On the tenth one, taken almost four decades later, I found Tombaugh's star again in outburst!

I now needed to examine every one of the 360 Corvus plates in the Harvard archives. Over the next three days I spotted ten more instances of the star exploding. Shortly after I returned home from Cambridge, I began monitoring the progress of this star through my own telescope. On March 23, 1990, I saw Tombaugh's star in outburst for the first time. (The star is far too faint for visual observation when it is not in outburst). Fifty-nine years to the day after its first recorded outburst, this star's light was for the first time being directly witnessed by human eyes.

TV Corvi is a pair of stars orbiting each other. The smaller one is a white dwarf probably no larger than the Earth. Over several months it "steals" hydrogen from its neighbor star. That companion star is probably only a few times bigger than Jupiter. (If it is that small, it is an example of a kind of body that is at the dividing line between a star and a planet.) Too big to be a planet but too small to undergo nuclear fusion, the other object is a brown dwarf. The two suns orbit each other in an area of space smaller than our own Sun.

There are other "cataclysmic variables," stars that vary explosively every few months. The explosion actually does no damage to the star, which many continue to blow off hydro-

VARIABLE STARS

The Pleiades. Gene and Carolyn Shoemaker and I took this photo of Messier 45 (the Pleiades) in 1989 using the 18" Schmidt Camera at Mount Palomar.

gen in this way thousands of times. SS Cygni, one of the best known, is a binary system whose outbursts have been followed by amateur observers since 1896. Although the eruptions occur every two months on the average, it is not known when this type of variable star will next have an explosion.

If a star can get brighter without warning, can it fade without warning? Although there are not many, a few such stars do exist. The most famous example is R Coronae Borealis, a star in the constellation of the Northern Crown, whose quiescent state is marked by a constant brightness level just at the limit of naked-eye visibility in a dark, country sky. Early in 1977 I watched it slowly fade until it was barely visible through my largest telescope. R Corona Borealis apparently suffers eruptions of very dark material into its atmosphere. Like soot from a furnace, this dark carbonaceous material thickens as it blots out the star's light, and it can keep the star faint for months at a time.

Not far from R is T Coronae Borealis, an explosive star that erupts not every few months but every century or so. Normally visible through a small telescope, in 1866 its brightness soared overnight till it rivaled the brightest star in the Northern Crown. In 1946 it repeated this performance. Most T Coronae Borealis observers watch the star every night, just in case the star undergoes a minor fluctuation in light, or just possibly, its first major outburst in many years.

The Orion Nebula, one of the most beautiful regions of the entire sky, is a stellar nursery filled with young blue stars, and a number of these stars change in brightness. Their brightness changes can be frequent and fast; in fact within just a few minutes a star can brighten by a fifth of a magnitude or more. Orion-type stars vary for two reasons. As a star's surrounding nebulosity changes in thickness, the star might slowly change in brightness. But real brightness changes are also going on within these stars.

After making more than ten thousand observations of these stars over some years, I grew familiar with their habits. Some stars, like T Orionis, vary slowly over long periods of time. But I have seen others, like V361 Orionis, appear to "flicker" as much as a fifth of a magnitude. Although the Orion stars are more accurately followed with modern CCDs, a visual observer can enjoy catching them as they show the carefree behavior of stellar youth.

REFERENCES

Houghton, Walter E., and Robert Stange G, eds. *Victorian Poetry and Poetics*. Boston: Houghton, Mifflin, 1968.

Williamson, Isabel K. "Delta Cephei instruction sheet."

CHAPTER 7

When Stars Die

> On découvrit cette année au mois d'octobre une nouvelle étoile près du pied précédent d'Ophiuchus. Elle étoit, dit-on, dès sa première apparition, plus belle que Jupiter, moins grande cependant que Vé: elle décrut ensouite graduellement. . . .
> —George Spate
>
> (A statement about the discovery by George Spate, on September 27, 1604, of a new star in the constellation of Ophiuchus. The star was brighter than Jupiter, but not as bright as Venus.)

The supernova we now know as Kepler's Star was not actually discovered by the famous astronomer, but it was he who did the major study on it at the time. According to A. G. Pingré's monumental review of 17th-century observational astronomy, the supernova was first reported by George Spate, a relatively unknown observer from Poméranie. He described a new star near the west, or preceding, foot of Ophiuchus, that was brighter than Jupiter but fainter than Venus. The star remained bright throughout 1605. It was a supernova, the result of a star whose central core has collapsed, sending the rest of its matter hurtling into space and shining as brightly as 100 billion suns.

This new star was an extraordinary event, the first in a

generation. But the previous one, Tycho's star in 1572, was almost unprecedented. Besides the supernovae of 1006 in the constellation of Lupus, 1054 in Taurus, 1572 in Cassiopeia, and 1604 in Ophiuchus, there was an eruption in Cassiopeia in the early 18th century. Because this latter star was obscured by our galaxy's foreground dust, it was not observed by anyone at the time.

These two stars fundamentally changed the way we look at the sky. Most stars, like our Sun, go through the relatively benign red giant phase before settling down as white dwarfs. We have seen how some of these stars undergo minor explosions in their shells. We have seen how some stars even undergo the core explosion called the helium flash. But for more massive stars, the end is utterly savage, resulting in the star's total destruction within a few minutes.

In the 1930s two astronomers, Fritz Zwicky and Walter Baade, working at the great observatories in California, developed the basic ideas for exploding suns that we use today. They were not studying the ordinary novae, where a star suffers an explosion that causes it to brighten by a hundred times, but the supercharged blasts that cause a star to shine, for a few days or weeks, with the intensity of hundreds of billions of suns. Such an event is called a *super*nova.

Type I Supernovae

In rare cases, it is possible for a star the mass of our Sun to become a supernova, but only if the star, in its white dwarf stage, happens to be one component of a double star. It may try to capture hydrogen from its neighbor. One consequence of this behavior can be a periodic nuclear explosion such as we saw in Tombaugh's star, or a blast every century, such as in T Coronae Borealis. But what if the captured material never ignites? If the dwarf keeps on gathering more and more matter, how massive can it become and still have the stability that degenerate matter has in a white dwarf?

Some decades ago, the Indian-born astrophysicist Subrahmanyan Chandrasekhar proposed that there is a limit of mass beyond which a white dwarf cannot stay a white dwarf, and that after that point, even its degenerate matter will start to contract, becoming even denser. As soon as that limit is reached, the star blows up. If the conditions are right, a white dwarf can end its life spectacularly after living its normal life as a star for many billions of years. It is even possible for a *single* white dwarf star to become a supernova, if its mass exceeds the limit of 1.4 times the mass of the Sun.

Type II Supernovae
There are apparently at least two versions of Type II supernovae, depending on how massive the star is. In either case, Type II explosions are the end result of stars that have lived too hard and too fast. Burning themselves out in just a few million years, compared to the 10-billion-year lifespan of a star like the Sun, these stars end their lives with great violence. In the weaker type, the star fuses its hydrogen, and then its helium, until the core is left with carbon. Like the helium flash in Sun-like stars, there is a *carbon detonation*, which occurs when all the carbon in the core ignites at once. This detonation may be strong enough to blow the core apart and lead to a supernova.

More massive stars survive carbon detonation, but in doing so they have bought only a few hundred years of time. Stars more than nine times the mass of the Sun are so hot that their carbon gradually ignites. A very massive star 25 times the Sun's mass, after spending seven million years fusing hydrogen to helium, fuses its helium supply into carbon in half a million years. The process of carbon ignitions begins gradually, so it is safely fused to oxygen in six hundred years. Then the oxygen fuses to silicon in the short time of about six months. At the end, in less than a day, the silicon fuses to form a core of iron.

If stars only understood nuclear physics, they'd know that an iron atom is so stable that it cannot fuse. Instead of releasing energy in a fusion reaction when enormous amounts of heat are applied, iron will absorb the energy. But since the star is unaware of that property, it tries to ignite its iron core anyway by contracting and heating it. Unable to keep up, the core suffers a final collapse. In less than two seconds the core crashes in on itself, carrying large amounts of still unused fuel; as the electrons crash into the nuclei of their atoms, they form neutrons and neutrinos, and a new kind of star we call a *neutron star*.

Around the collapsing stellar core, bedlam reigns. A shock wave pushes the star's outer layers away at tremendous speed. In the titanic explosion, the star outshines the combined light of all the stars in its galaxy.

The Supernova of 1987
Until just a few years ago, what we knew about supernovae was limited to the examples we could find in distant galaxies. Since the invention of the telescope, no supernova had appeared in our galaxy or its two nearest neighbors, the Magellanic clouds. It had also been more than 100 years since the last near supernova eruption in the Andromeda galaxy, which occurred in 1885. On the morning of February 22, 1987, Ian Shelton, a Canadian astronomer at the University of Toronto's telescope at Las Campanas in Chile, took the first plates of a new patrol program. His goal was to discover new variable stars in the Magellanic Clouds. Two mornings later, he had completed the exposures and developments of the first three plates of the larger of the two clouds. Although the first two plates showed the Large Magellanic Cloud (LMC) as it should be, the third contained a new bright star. Thinking that something so bright must be a defect in the photographic process, he went outside to check the LMC himself. It was no defect. Staring at him from 169,000 light years away was an exploding star!

Two other observers independently found the supernova. At a nearby telescope at the same observatory, night assistant Oscar Duhalde was making a routine check of the night sky when he also noticed the new star. The third discoverer was across the world in Nelson, New Zealand. Albert Jones, an experienced variable star observer, noticed the supernova as he checked the sky. "I was observing stars in the south when I noticed clouds coming over from the west," Jones recalls. Since the Large Magellanic Cloud, our neighbor galaxy some 160,000 light years away, contained some variable stars he wanted to observe, he quickly swung his telescope in its direction. "I looked in the finder to locate the variables near the Tarantula Nebula but was surprised to find a bright blue star there. There was the Tarantula Nebula, so I knew I had the telescope in the right place." (Jones, 2000) Clouds quickly covered the intruder, but Jones hurried to call Frank Bateson, director of the Variable Star Section of the Royal Astronomical Society of New Zealand, to inform him of his find. After the clouds cleared hours later, Jones observed the star again, realizing it was bright enough to see with the naked eye.

The story next focuses on the cramped offices of the Central Bureau for Astronomical Telegrams. Operating out of the Harvard-Smithsonian Center for Astrophysics in Cambridge, Mass, Daniel Green and Brian Marsden filter through anything new in the sky that moves or explodes. Just before 9 A.M. on the morning of February 24, the telex machine crackled with the first discovery message of the supernova. A few minutes later, a telephone call from Chile confirmed that astronomers Ian Shelton and Oscar Duhalde had independently found the first bright supernova in over 380 years. The independent report from Jones arrived later.

By late morning telephones were ringing incessantly as Marsden completed his announcement circular. Then the first spectroscopic observations came as night fell on the South African Astronomical Observatory. These data confirmed that the bright new star was indeed a supernova. By the end of the day in Cambridge, the announcement was published and distributed around the world:

> SUPERNOVA 1987A IN THE LARGE
> MAGELLANIC CLOUD
>
> W. Kunkel and B. Madore, Las Campanas Observatory, report the discovery by Ian Shelton, University of Toronto Las Campanas Station, of a mag 5 object, ostensibly a supernova, in the Large Magellanic Cloud at R.A. = 5h35m.4, Decl. = 69 16' (equinox 1987.2), 18' west and 10' south of 30 Dor and possibly involved with the association NGC 2044. The discovery was made around Feb. 24.23 UT on a 3-hr exposure with a 0.25-m astrograph beginning on Feb. 24.06, and the object had evidently brightened by at least about 8 mag since the previous night.

We learn so much in the early moments of a supernova eruption, for it is in that early moment that a star evolves to its limit of nuclear fusion and collapses that all of the heavier elements necessary for life, like oxygen and carbon, are released into space. One beautiful lesson came a few hours *before* the supernova's light reached us; bursts of neutrinos completed their 169,000 year journey from the collapsing star to Earth, and were captured by four neutrino detectors buried deep underground.

Since you began reading this sentence, some 100 billion neutrinos have forced their way through your body. Because they have virtually no mass, these particles fly through almost anything. Most of the neutrinos in our area come from the Sun, but some come from other stars. As the iron core of the dying star in the Large Magellanic Cloud began its fateful, two-second collapse, immense numbers of neutrinos cascaded away. It was fascinating to see theory turned into practice when at least twenty-five neutrinos completed their long journey and were detected here. It is virtually certain that some neutrinos from the distant dying sun passed through the bodies of everyone on Earth.

The progenitor star (the star before it became a supernova) had an unusual history in the few thousand years before its death. Some 20 times the mass of the Sun, it had possibly grown into a red giant. Several thousand years ago it suffered some kind of explosion in which it lost its outer shell into a cloud. With less mass, it shrunk and heated, changing into a blue giant star to live out the last, brief phase of life before its final collapse. In 1998 the Hubble Space Telescope recorded the old cloud brightening as it suffered the impact of energy from the expanding supernova remnant.

Remnants of Supernovae
Supernova remnants exist throughout the sky; these wispy clouds tell of stars that blew up many thousands of years ago. Their remnants are still hot, still expanding through space. Most famous is the Crab Nebula—so called because it resembles a ghostly crab—the remains of the mighty explosion that so many people around the world witnessed in 1054. It also holds the honor of being he first entry in a catalogue of comet-like objects made by the French astronomer Charles Messier, more than two hundred years ago. He labeled the Crab Nebula M1. In only 700 years, the explosion produced a shell of gas bright enough that Messier could easily see it through his small telescope. In 1969, a team of astronomers using the Steward Observatory telescope at Kitt Peak, Arizona, found an object at its center, a pulsar that spins 30 times per second.

WHEN STARS DIE

A Song from a Supernova

We have seen how vital supernova explosions are for the development of life—it is in the explosion of a star that organic elements are spread throughout the galaxy and become part of the population of comets that, in turn, bring them to planets as the building blocks of life. In the following song, written by my friend Peter Jedicke, the concept of life from death takes musical root. (Sung to the tune of *When Johnny Comes Marching Home*.)

> The stars go nova one by one, KABOOM! KABOOM!
> Nucleosynthesis is done. KABOOM, KABOOM!
> The supernovae dissipate
> What fusion energy helped create,
> And the stars go nova in the galaxy.
>
> The heavy elements are born, KABOOM! KABOOM!
> And from the stellar cores are torn, KABOOM! KABOOM!
> Shells of gas are strewn through space,
> Distributing matter all over the place,
> And the spiral arms are littered with debris.
>
> As years go by the remnants spread, KABOOM! KABOOM!
> But the Universe is far from dead, KABOOM! KABOOM!
> To eliminate the tedium,
> The interstellar medium
> Forms the molecules that make up you and me.

REFERENCES
Jones, Albert. 16 December, 2000.

CHAPTER 8

The Undiscover'd Country

> . . . who would fardels bear,
> To grunt and sweat under a weary life
> But that the dread of something after death,
> The undiscovr'd country from whose bourn
> No traveler returns, puzzles the will
> And makes us rather bear those ills we have
> Than fly to others that we know not of?
> —William Shakespeare,
> *Hamlet*, act 1, scene 3

It would be quite a privilege to have a front-row seat at the death of a massive star. We don't get to see all that is going on because the explosion that rips away the star's outer layers blocks our view of its heart. If the progenitor star was below a certain mass, the core's collapse would stop with the formation of a neutron star the size of a town, made of neutrons and sending out brilliant beams of energy every second. In more massive stars, the core's collapse would continue until it had shrunk to an unseen point of immense gravity that, like the Cheshire cat's grin, is all that's left of what once was a beautiful star.

Pulsars

When my telephone rang on the cold winter evening of January 5, 1968, I expected it to be a routine phone call from Thomas Meyer, my high school friend. But he had news. "I just heard," he said, "that a radio telescope has picked up ordered signals from space." We thought that this must mean that life *does* exist elsewhere, that we are not alone. Indeed, some radio telescope had picked up strange signals in the preceding few months. The astronomer, Antony Hewish of Cambridge University, was doing a survey of "interplanetary scintillation," a rapid change in signal intensity caused by the "solar wind" of radiation coming from the Sun. The effect of this scintillation is that galaxies observed at radio wavelengths seem to show variations in their intensity. Jocelyn Bell Burnell, a student of Hewish's, helped build the radio telescope and was carrying out its observations on the evening of August 6, 1967, when she noticed an unusual signal. After following it for some months, they suddenly received a stronger version; Burnell was shocked to find that the signals repeated at one second intervals. More accurate timing revealed that they were repeating to an accuracy of better than a millionth of a second! Thinking that these might be signals from a distant intelligent civilization, she gave the signal the acronym LGM, for Little Green Men.

The signals Burnell discovered turned out not to be artificial. She and Hewish quickly learned that the signals were not coming from a world circling a star, but from a star itself. Also, the signals were sent over a very broad range; an intelligent civilization trying to communicate with other beings would use a single frequency, like a radio station, and not waste its transmitter over many frequencies.

Even though there were not little green men, the find was sensational. It was the first detection of a star formed out of a supernova, a pulsar or neutron star rotating very fast, and sending out radiation in a focused beam like a lighthouse.

COSMOLOGY 101

The object forms in a fraction of a second as a star's iron core implodes just before the star's outer layers are blown into space in the supernova explosion.

Neutron stars are spheres about 12 miles across—the size of a typical comet! The star's outer shell is a skin as hard as iron. As "degenerate matter," electrons move randomly throughout the material just below the crust. Deeper in the star, the electrons and protons are gone; there is only a very dense fluid of neutrons. It is possible that the star's core is made of subatomic particles called quarks. The whole star is incredibly dense; a teaspoonful of neutron star has the mass of a mountain.

Several hundred pulsars have been spotted, pulsing at rates as high as 1500 times per second! Since the pulsars flash twice each rotation, these 12-mile bodies must be spinning at rates of several hundred times per second. Two pulsars have been spotted in visible wavelengths; all the rest reveal themselves only with radio telescopes. One of these two is at the center of an old supernova remnant in the constellation of Vela, and the other is the spinning, pulsing core of the great supernova of 1054.

Although the stellar ghosts we call pulsars rotate very fast, their rates slow down over long periods of time. More interesting, the Vela pulsar has "glitches" in which, every few years, the star's crust breaks slightly and then reforms very slightly closer to the star's center. This "starquake" effect occurs when the star's regular pulsing suddenly speeds up for a brief time, by a fraction of a millionth of a second, then returns to its previous rate. Starquakes are common in some pulsars.

The discovery of pulsars is important for another reason. These distant objects are lighthouses in the sky, their flashing beacons providing a reference point for light years around. Since most pulsars are found in our galaxy's spiral arms, mapping their locations provided a new way of tracing the extent of the arms.

Black Holes

The core of a collapsed star with less than 1.4 times the mass of the Sun ends up as a white dwarf, and the core of a star up to 3 times the Sun's mass forms a neutron star. But what if a very massive star, with 5 to 50 times the mass of our sun, still leaves more than three solar masses of material after its supernova? Or what if the star collapsed right through the neutron star stage, without a supernova explosion? It would continue to collapse indefinitely until it became as small as a city, a house, a doorknob, and then disappear altogether, leaving only its gravity—and so much gravity that even light could not escape from it.

There is no force in the Universe that can stop the collapse of the core of a star with more than three times the mass of the Sun. As the core contracts, its own radiation is gravitationally bent as it tries to escape into space. After the star has collapsed to a certain size, nothing can escape from it. If you stand on the theoretical surface of a black hole and shine a flashlight beam into the sky, the beam would return to you. If you travel past a black hole, you are safe from its clutches so long as you keep away from its *event horizon*, the spherical area around the black hole within which nothing can escape. If you take a chance and "drop in" feet first, you would suffer quite a fate. In less than a hundredth of a second, your feet would go in more strongly than your head and would separate. The other parts of your body would separate in the same way, the parts closer to the center of the black hole being pulled much faster than those farther away. Then, what's left of your body would devolve into individual atoms, then your atoms would break up into separate electrons, protons, and neutrons, and then each particle would stretch and be broken into some new particle we are not even aware of.

Inside a black hole, nothing would be stationary—everything would continue to fall toward the center. According to Einstein's theory, nothing can stop the collapse until all the

This page and opposite: The Crab Nebula. One of the two stars in the middle of this nebula is the pulsar remnant of the supernova that created the nebula in 1054. (Courtesy NASA/HST and Paul Scowsen.)

mass is in a single point, called a singularity. However, some theorists believe that a true singularity is impossible, and that some quantum mechanical effect eventually would stop the collapse.

HDE 226868

Being right in the middle of the Milky Way, the constellation of Cygnus has lots of faint stars that we see only as the Milky Way's band of hazy light. One of these stars has the unremarkable name of HDE 226868. It is a blue supergiant, a hot and rare O-class star that is some 15 times more massive

THE UNDISCOVER'D COUNTRY

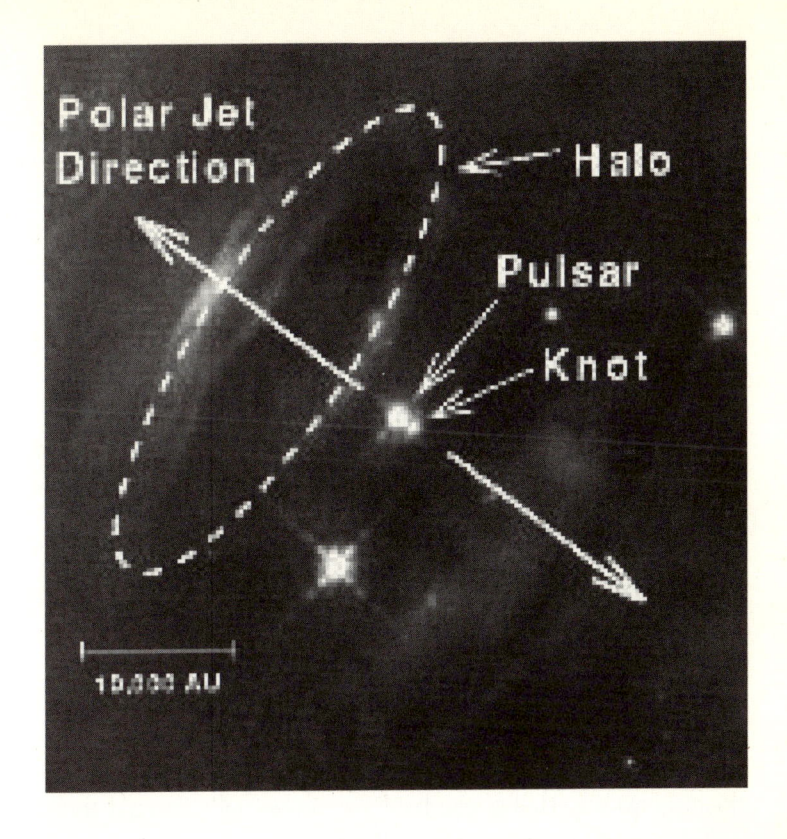

Polar Jet Direction
Halo
Pulsar
Knot
15,000 AU

than the Sun. The star first came to astronomers' attention when they discovered it at the center of a strong radio emission source called Cygnus X-1. Something at that position was rotating very quickly, once every few thousandths of a second. But no lumbering blue supergiant can rotate anywhere close to once a millisecond. Only pulsars—the rotating neutron stars that are the remains of a stellar core after a star has exploded as a supernova—vary that fast. Could the blue star be orbiting one? Indeed. When astronomers calculated the blue giant's orbit they discovered just how massive this companion was. Although the star and the unseen object are orbiting each other, the invisible one is more than a quarter—and possibly as much as half—as massive as the bright blue supergiant! Too massive to be a neutron star, the unseen object is suspected to be a black hole.

In the constellation of Monoceros is a variable K-class star called V616 Monocerotis. It is orbiting a compact mass, called A0620-00, which emits powerful X-rays; the little object is at least 3.2 times the mass of the Sun, and is the strongest case yet for the existence of a black hole.

Einstein-Rosen Bridges
Is it possible to have two black holes that formed simultaneously, without the need to be the result of a star's collapse, and can the interior of one be equivalent to the interior of the second? With physicist Nathan Rosen, Albert Einstein suggested that it is mathematically possible for a black hole and its opposite, a "white hole" in another part of the galaxy, to be joined by a *wormhole*. It is also theoretically possible that the exit of the wormhole would be in a different universe that would have no relationship in either space or time to ours. This "Einstein-Rosen bridge" would be a convenient shortcut and has been used in science fiction. However, there is no way that these bridges can actually form if they are not present at the birth of the Universe, and they would not have the stability that space travelers demand of them. In mathematics they can be predicted, but there is no astrophysical reason to believe they actually exist.

Like the smile of Alice's vanished Cheshire cat, the ultimate end of a massive star offers the chance to think about the night sky in ways that stretch our imagination. It is a good thing that we can envision all this from the safety of our own world, as we orbit a stable, and small, G-class star.

Neutron Star Song
You'd never expect that the complex physics behind a neutron star could be reconciled to the tune of Loretta Lynn's country song *Jealous Heart*. But from the pen of Canadian

physics teacher and amateur astronomer Peter Jedicke comes one of the most intelligent astronomical songs ever written.

The song includes ideas and principles discussed in this book—the Hertzsprung-Russell diagram, and a star's thermonuclear reactions depleting its hydrogen. The chromosphere is an inner layer of a star's atmosphere. The verse about the magnetic lines of force is a musical explanation of the fairly difficult astrophysical concept of a synchrotron emission source, in which electrons race out along magnetic fields at velocities approaching the speed of light. "The important thing behind any strange emission," Jedicke explains, "is that it has to come from somewhere." In an extremely dense neutron star electrons are being moved about in unusual ways, traveling around lines of magnetic fields like a helix. How does one get a concept like that into a country song? With a little knowledge of how stars work, and a lot of humor and fun.

> Neutron star, oh neutron star, you're massive
> And your tidal forces are intense,
> You have crushed your atom shells to pieces;
> Neutron star! your gravity's immense.
>
> You were once a star like all the others,
> Shining brightly in the evening sky,
> Till your thermonuclear reactions
> Consumed all your hydrogen supply.
> Neutron star, oh neutron star, you're spinning
> Round and round at such a fever pitch.
> You conserve your angular momentum
> And speed up at every little glitch.
>
> You were once a star like all the others
> Somewhere on the Hertzsprung-Russell graph;

Now you're in the lower left-hand corner,
Stellar mass reduced by more than half.

Neutron star, oh neutron star, you're pulsing—
Twisting your magnetic lines of force,
And electrons spewing from your axis
Form a synchrotron emission source.

You were once a star like all the others
Till your hydrostatic balance failed,
And you lost your radiation pressure,
And your outer chromosphere exhaled.

REFERENCES

Bok, Bart J. "Summary and Outlook: Symposium on the Spiral Structure of our Galaxy." Basel University, September 4, 1969.

Hewish, Antony. "Pulsars." *The International Encyclopedia of Astronomy*. Ed. Patrick Moore. New York: Orion, 1987, 322-324.

PART THREE

OF GALAXIES AND COSMOLOGY

CHAPTER 9

The Great Clusters

> Many a night I saw the Pleiads, rising thro' the mellow shade,
> Glitter like a swarm of fire-flies tangled in a silver braid.
> —Tennyson, *Locksley Hall*

Star Night! Just saying those words were exciting back in the 1960s when the Royal Astronomical Society of Canada's Montreal Center put on its annual Star Night in Westmount Park. It was a chance to show the community the night sky and the work of the RASC; the hope was that huge numbers of people would line up at some two dozen telescopes, pointed at different objects. The evening of Wednesday, October 14, 1964, one of those Star Nights, was to be a highlight of my observing life. Each telescope operator had been assigned a specific object for the evening. When I learned that my assignment was an object called Messier 15, I adopted the distant cluster as a close friend. Probing into books and atlases, I learned as much as I could about that distant crowd of stars in the autumn night.

Messier 15, it turned out, is a globular star cluster, an enormous grouping of more than fifty thousand stars, all peacefully circling the center of our own galaxy. It is 34,000 light years away from us. I learned that this cluster is very old, formed long before our solar system was even a cloud of hydrogen atoms. I found myself wondering what it would be like to set up housekeeping on a world orbiting a sun in that cluster. Every night its light would dominate our sky. With as many as 30 stars in every cubic light year, stars at the center would be separated by no more than a few light months instead of the 4 years in our corner of the galaxy. Filled with stars, the sky would never get darker than a late twilight sky here at home.

Later I also imagined what it would be like if a planet circled a star at the *edge* of a distant globular cluster. Far from the center of our galaxy, this world's sky would be sparser than ours, containing fewer stars. However, taking up fully a quarter of the sky would be the vast maelstrom of the cluster itself, brightening up the night sky. What would a civilization think of the giant orb in the night? How would such a sight affect their perception of nature, or their religious beliefs? Would they think that Heaven was the center of the cluster?

Back on Earth that autumn Wednesday night, clouds blocked the sky until about 9:30 P.M. Then, for the first time, I treated a large group of people to a look at the distant cluster in deep space. Most had never even looked through a telescope before, and here I was, showing them a mighty cluster of stars at the very outskirts of our galaxy. Every time I look at M15 my fond memories of that night are renewed.

Globular Clusters

Before Abraham Ihle looked up into the sky in 1665 and discovered the fuzzy patch of light in the constellation of Sagittarius we now call Messier 22, globular clusters were not a part of our known universe. In 1677, Edmond Halley, while observing from St. Helena, was fascinated by the large oval-shaped globular cluster Omega Centauri. In 1714 Halley recorded the great cluster in Hercules. By the end of the 18th century the British astronomer William Herschel began his great survey of the northern sky, his goal being to catalogue everything his telescopes could see. He described some objects—the ones that turned out to be globular clusters—as shining with "a mottled kind of light," which later and better telescopes resolved into the many stars that make up a cluster. One of the most recent finds took place in 1932, when the intrepid observer Clyde Tombaugh discovered that the object NGC 5694 was in fact a globular cluster shining at us from the incredibly great distance of 105,000 light years, on the other side of our galaxy.

Globular clusters contain many thousands of stars. Masses of stars that typically lie on the outskirts of our galaxy, the globular clusters are among the best objects to view, even from a city sky. We know of 131 globular clusters in our galaxy, and there are probably another hundred whose light is blocked by the great intervening clouds of dust in space. While some clusters exist near the center of the galaxy, others live out their long lives in the lonely parts of space far away from the density of the galactic center. One of them, called NGC 2419, lies in the constellation of Lynx, just north of the bright star Castor in Gemini. It first revealed itself as a globular cluster to the sharp eyes of Carl Lampland at Lowell Observatory. This cluster is so far out of our galaxy—more than three hundred thousand light years away—that Harlow Shapley thought it to be totally free of its gravitational pull, and he called this cluster an "intergalactic tramp." The cluster, it turns out, *is* probably connected to our galaxy, looping around its center in a very eccentric orbit.

Because the globular clusters are concentrated in the half of the sky closest to the center of the galaxy, the summer sky,

THE GREAT CLUSTERS

with its brilliant Milky Way, offers most of the clusters. By contrast, the winter sky offers the diminutive Messier 79, south of Orion's belt. Messier 79 is 43,000 light years away.

Although globular clusters tend to have a similar appearance based on the enormous gravitational pull of their stars, they do come in a variety of appearances depending upon how closely their stars are concentrated. Shapley classified them in a I to XII scale, where a type I cluster is densely concentrated, and a XII has virtually no concentration at all, its stars being loosely distributed. The full Shapley scale is as follows:

- I: high concentration toward center
- II: dense central condensation
- III: strong inner core of stars
- IV: intermediate rich concentration
- V: intermediate concentration
- VI: intermediate
- VII: intermediate
- VIII: rather loosely concentrated toward center
- IX: loose towards center
- X: loose
- XI: very loose towards center
- XII: almost no concentration in center

The stars that form globular clusters are quite different from ordinary stars. The brightest are yellow or red giants. The clusters are also immensely old—some calculations show that Messier 13 has been around for some fourteen billion years. If that figure is correct, globular clusters are among the oldest structures in the Universe. Globular cluster stars tend to be devoid of heavy elements, and they are not surrounded by gas and dust. Hence the stars in the clusters must have been formed before the gas surrounding them was enriched by heavy elements from supernovae.

There is another difference, albeit a subtle one, between globular clusters existing within the galaxy and clusters that orbit in the outlying areas of our galaxy. The remote clusters tend to be smaller because, as they orbit the galaxy, they will occasionally pass through dense layers of dust that disrupt them. The result: The pressure of the dust can be strong enough to push some stars out of their clusters entirely! The area around our galaxy, called the galactic halo, is filled with old, lone stars that used to belong to globular clusters.

Open Clusters

Perhaps you are familiar with "The Seven Sisters," a grouping of stars also known as the Pleiades. The Pleiades is a star cluster, but with far fewer stars than the mighty membership of a globular like Messier 13. When I first looked at the Pleiades through a telescope, I was impressed by how many stars it contained! Far more than the seven stars visible with the naked eye, I counted several dozen. Actually, the cluster contains some 500 stars, only one or two of which are still surrounded by the remnants of the clouds out of which they were born.

Just south of the Pleiades is a bigger cluster called the Hyades. It appears bigger because it is much closer to us than the Pleiades cluster is; in fact, the Hyades cluster is our closest visible star cluster. Such clusters are called "open" or "galactic" clusters, since they are all found within the galaxy rather than circling it as do some of the globulars. However, there is a closer cluster. The Sun is actually passing through the remains of a loosely knit star cluster whose center is about 75 light years away from us. This "Ursa Major group" consists of stars like Beta, Gamma, Delta, Epsilon and Zeta in Ursa Major—most of the stars of the Big Dipper—and possibly Alpha Coronae Borealis, the brightest star in the constellation of the Northern Crown. Moving through space in almost the same direction is a stream of material that includes stars as far afield as Sirius, Alpha Ophiuchi, Delta Leonis, and Beta Aurigae. Just what the relation is, if any, between the cluster and its stream is unknown.

How can we be certain that a cluster of stars is real—its members traveling through space together—instead of a chance optical alignment of stars? The answer brings us back to Clyde Tombaugh, the discoverer of the distant globular cluster NGC 5694. He found that cluster as part of the planet survey that began in 1929, which led to the discovery of Pluto in 1930. The search strategy involved taking many photographic plates of the sky. Each photograph was repeated twice, and Tombaugh placed the best two pictures into a device called a blink microscope. (As mentioned earlier, it was by scanning these pairs of plates for moving objects that he discovered Pluto.)

By 1945, Tombaugh had amassed a collection of plates so vast that they covered almost all of the sky visible from the observatory site near Flagstaff, Arizona. The plates contained images of some 45 million stars. About ten years later, Henry Giclas launched a new program that involved rephotographing the sky, this time to search each one of those 45 million to find out how many had moved slightly, in which direction, and by how much. A star's real movement across the sky (not its apparent moving because of the Earth's rotation) is called its proper motion, and the Lowell Observatory survey succeeded in cataloguing the proper motions of many stars. The idea that stars actually moved relative to one another was first noted by Edmond Halley in 1718. Studying the 1200-year old star charts that Claudius Ptolemy had prepared, Halley noticed that two of the sky's brightest stars, Sirius and Arcturus, had changed their positions substantially over that time.

If a group of stars appears to be a cluster, astronomers investigate to see if its stars have the same proper motion. It is also important to know whether the group is moving toward us or away from us, and at what speed; we call that motion radial velocity. If a group of stars share the same proper motions and radial velocities, then it is said to be a true open cluster. The Hyades, for example, is a collection of stars moving through space. From Earth, however, the cluster is punctuated by a red star named Aldebaran. Although that bright star adds to the cluster's beauty, it is moving in a different direction and is not a true member of the cluster.

In the early 20th century Harlow Shapley classified open clusters according to how concentrated their stars were, using the letters a through g:

> a: barely recognizable associations
> b: vaguely defined groupings
> c: very loose and irregular
> d: loose and poorly concentrated
> e: intermediate rich
> f: fairly rich with stars
> g: considerably rich and concentrated

T Associations

When the Sun was born, the giant molecular cloud out of which it was formed contained material for much more than a single solar system. Other stars, maybe dozens, maybe hundreds, were formed at the same time. The grouping was an open star cluster that might have looked like the grouping at the center of the sword of Orion, whose beautiful nebula contains stars only a few million years old. Cosmologically, such stars are very young. Another association lies in Sagittarius, whose Lagoon nebula contains a cluster of many young stars.

These very young star clusters are called T associations. Named for the young star T Tauri, these associations were first pointed out by the Russian astronomer Viktor Ambartsumian.

THE GREAT CLUSTERS

T associations are typically short lived, not longer than 500 million years.

Past and Future
What would an ancient stargazer have thought while gazing into the night toward the constellation of Hercules, and spotting a strange, misty spot there? With the sky full of solid points of light, the little fuzzy speck might have caused him or her to wonder. To one of the ancients, the fuzzy light was barely visible.

Now we move forward to November 16, 1974. At Arecibo, Puerto Rico, a newly built radio telescope was about to begin its career of studying the heavens. On this special night, the telescope would not receive a signal but send one. A carefully constructed message telling others about ourselves had been prepared, and at the appointed hour, the telescope pointed to the same misty spot seen by the ancient stargazer. Although the thought was different—the people running the telescope knew that the misty spot was a globular star cluster—the wonder was still there, still the same as it was so many generations ago. The message silently began its long journey that night, and at the speed of light it will take more than 23,000 years to reach the great Hercules cluster. Will it reach a planet whose sun is on the cluster's perimeter, or is the recipient to be the world at the cluster's center? And what will this civilization-to-come think of the people so far away who sent the message? These ideas are worth thinking about as we turn our telescopes to marvel at the wonder of the globular cluster in Hercules.

A Sampler of Clusters
Messier 13: The grandest cluster in the northern sky. Deep in the constellation of Hercules, this cluster is faintly visible to the unaided eye. Through even the smallest telescope, the

outer parts of the cluster might be distinguished into individual stars. Shapley class V.

Omega Centauri: Although this beautiful cluster lies deep in the southern hemisphere, it is visible from the southernmost latitudes of the United States. The cluster has such a decidedly oval appearance that some astronomers think it might be two clusters that somehow joined together. Shapley class VIII.

Messier 15: A beautiful cluster in the autumn sky, this object is easy to find near the star Epsilon Pegasi, and although, at Shapley Class IV, it is fairly concentrated in appearance, a small telescope will resolve it into stars.

Pleiades: A beautiful open cluster in Taurus, it is easily visible to the naked eye even from a bright city sky. Class c.

Hyades: Also in Taurus, this open cluster spreads out over a much larger area of sky than the Pleiades. Class c.

REFERENCES
Tennyson, Alfred. "Locksley Hall." *Victorian Poetry and Poetics*. Ed. Houghton and Stange. n.p.: n.p., n.d. n. pag.
Thoreau, Henry. *Walden*. New York: Peebles P International, n.d.

CHAPTER 10

The Milky Way

A host, of golden daffodils;
Beside the lake, beneath the trees,
Fluttering and dancing in the breeze.
Continuous as the stars that shine
And twinkle on the milky way,
They stretched in never-ending line
Along the margin of a bay. . . .
—William Wordsworth, 1806

Of all the objects of sheer beauty in the night sky, the Great Sagittarius Star Cloud is one of my favorites. It is also one of the most ignored. Many times at the Texas "Star Party," dozens of telescopes would be set up on distant galaxies and clusters of stars. I have heard people exclaim loudly through the night about the structure of some dust lane in a faint galaxy, while all alone, the Great Sagittarius Cloud shone majestically over the observing field, its light almost brilliant enough to cast a shadow, but quite ignored by almost everyone.

The Great Sagittarius Star Cloud (GSSC) even foiled Messier, which is a bit surprising because he did remember its smaller, northern neighbor as No. 24 in his catalogue. Messier 24 is not really a star cluster, as Steve O'Meara records: "Commonly called the Small Sagittarius Star Cloud, M24 is a virtual carpet of stellar jewels, laid out across 330 light years of space." Observing through his telescope, O'Meara reports that "no sight in the visible universe shares M24's mystical qualities." (You'll see much more on the French observer Charles Messier and his famous catalogue in the next chapter.)

I agree with O'Meara's assessment of M24's beauty—unparalleled until you head south to observe the Great Sagittarius Star Cloud. For something so beautiful, there is precious little written about it. I think that the best description comes from the late Robert Burnham Jr. whose *Celestial Handbook* (1978) describes it and its surroundings in 25 pages, a lot of which is poetry. The GSSC is, after all, not an object but a journey—not a deep-sky object to be added to an observing list but an experience to be treasured. The GSSC is the closest thing the deep sky has to wandering across the Moon with a telescope, crawling into craters, climbing over mountain ranges, and sliding down the Straight Wall. I could not believe that walking 100 feet to my 16-inch telescope and pointing it to the GSSC would launch me on a pilgrimage through star-studded fields interlaced with meandering rivers of dark nebulae.

How often on a clear night have our eyes been caught by that spattering of light across the sky that we call the Milky Way? We have explored stars, and clusters of stars; now we look at the home of all these suns, the Milky Way Galaxy. The Milky Way crosses the sky as a huge, irregularly shaped circle. When we gaze at it, we are looking *into* our galaxy. The band is thickest in Sagittarius, where the GSSC is; the galactic center is slightly to the southwest, near the Sagittarius-Scorpius border. One would suppose that because we live in this galaxy, we would understand its shape and appearance. But that understanding is similar to trying to figure out what a house looks like when we've only seen a room or two from the inside. To begin, we need to look out the window and see how neighboring homes appear.

Before Vesto M. Slipher, astronomer at Arizona's Lowell Observatory, did a small experiment almost a century ago, we had no idea about the appearance of the single basic family structure of our Universe, a galaxy. Lowell was and is a unique place. Founded by Percival Lowell of the famous Lowell family of Boston, the observatory's main goal was to study Mars. But Lowell had an idea about planets around other suns as well, and he thought that certain fuzzy patches in the sky called "spiral nebulae" could be stars with new planetary systems starting to form around them.

To find out, Lowell asked his friend Slipher to photograph these nebulae under the light of a spectroscope. It was a challenging project. Using the observatory's long and majestic 24-inch refractor, Slipher spent as much as two full nights taking a single exposure to gather enough light to reveal the spectrum of a single nebula. Slipher's results just added to the mystery of the spiral nebulae—the entire spectra of these nebulae were shifted toward the red end by various amounts. Because of a lack of telescope power, he was unable to interpret these results, and their mystery remained for almost a quarter century. While Slipher used a 24-inch diameter refractor at the turn of the century, by 1924 Edwin Hubble had the great 100-inch reflector at Mount Wilson at his disposal. Hubble and Milton Humason took such clear photographs of nearby spiral nebulae that they could resolve them into individual stars. These nebulae were hardly solar systems; they were galaxies like the Milky Way.

Detecting Spiral Structure in Our Galaxy

Before WW II, most astronomers thought of our galaxy as being generically round with no particular structure. But during the war years, astronomer Walter Baade, also at Mount Wilson, was able to take advantage of the unusually dark sky that was a consequence of the wartime blackout in Los Angeles to undertake a long series of observations of nearby spiral galaxies. Baade studied their spiral patterns, and detected two kinds of stars, which he called "populations." Population I stars, like the Sun, are rich in what we call metals. In astronomy, a metal is defined as an atom heavier than helium. Metal-rich stars tend to be found in the disk and spiral arms of the galaxy. Population II stars lack these metals, and tend to be found in a galaxy's center, or in the globular clusters, and in the halo that surrounds a galaxy. Baade also identified regions of hydrogen gas that characterize the spiral arms, and then proposed that it should be possible to detect spiral structure in our own galaxy by working the science in reverse: If someone traced areas of Population I stars, especially hot O and B stars with their associated hydrogen nebulosity, a spiral shape to our own galaxy might be detected.

In 1951, a team of astronomers led by William Morgan at Yerkes Observatory, followed up Baade's idea. By studying the distribution of stars of different populations in the solar neighborhood, they detected evidence of two spiral arms, which they called the Orion and Perseus arms, plus part of a third called the Sagittarius arm (Sky and Telescope, 1952). But Morgan's initial work detected spiral structure to the relatively short distance of 15,000 light years. To verify that these arms were real, astronomers needed a way to probe at least five times farther. Walter Baade's wartime research had also found dense regions of molecular hydrogen, called $H2$ regions, that seemed to outline the spiral structure of other galaxies. These regions are difficult to detect in optical wavelengths, but a new kind of science called radio astronomy was gaining a foothold; using large antennae, astronomers could listen to the sky at radio frequencies, as well as look at it through optical telescopes. The radio telescopes offered a totally new way to confirm the idea that our galaxy is a spiral. Using radio telescopes, astronomers detected hydrogen's distinctive 21-cm emission band to far greater distances.

With these data from radio telescopes, our concept of the basic form to our galaxy began to take shape by the early 1960s. Surrounding a central bulge are three spiral arms, in the directions of Orion, Perseus, and from Sagittarius to Carina. Because of obscuring matter between the stars, it was still hard to tell whether a particular extension that might be an arm is a major arm of our galaxy, or simply a small spur of a larger arm, most of which is hidden from view. Like trying to map the shape of a forest when all you can see clearly is a cluster of nearby trees, galactic mappers have a good view only of the part of our galaxy nearest us. For many years, for example, the astronomer Bart Bok suspected that one arm stretched through the northern constellation of Cygnus to the southern constellation Carina. However, more recent investigations detected an arm in the directions of Orion and Cygnus, a second arm in Perseus and part of an arm in Sagittarius and Carina. In a sense this Carina-Cygnus arm was demoted, since only the southern half exists.

By the late 1970s, the galaxy was discovered to be larger than previously thought. Earlier estimated to be half the size of our neighbor Messier 31, the Andromeda galaxy, our home galaxy is now believed to rival Andromeda in size. Although the existence of our galaxy's vast outer regions was suspected by that time, it was not until the 1980s that the extent of the galaxy's outlying areas was understood. We now have detected several components of our galaxy. The galaxy's main disk is filled with Population I stars, which includes the center and the spiral arms. There is a spheroidal bulge around the nucleus as well as a halo of thinly distributed stars around the disk, (population II stars) and a huge "corona" of material that is so vast that it couples the Milky Way with other large structures, like some distant globular clusters, the Magellanic Clouds, and seven dwarf galaxies. Until the late '60s, most scientists thought that the Milky Way had a mass equivalent to 200 billion suns. But if the galaxy's corona extends as far as 900,000 light years, then the number of stars in the galaxy could be as many as 400 billion.

The Center of the Galaxy

In the mid-1950s, a bright radio source (though optically invisible) called Sagittarius A was proposed as the site of the center of our galaxy. More recent observations have moved the center slightly to another very dense radio source called Sagittarius A* (A prime). The galactic center is not far in the sky from NGC 6451, an open cluster not related to the center but less than two degrees southeast from it. This cluster is a wonderful object to show at star parties as the deep-sky object closest to the center of our galaxy. Its position is RA 17h 50.7m, and Dec -30 degrees 13m. Thanks to the presence of intervening dust clouds, the center is in a relatively sparse section of sky at RA 17h 45.6m and Dec -28 degrees 56m.

Some 26,100 light years away, Sagittarius A* has received intensive study in recent years. The radio source and its surrounding stars cannot be seen in visible light, because of obscuring dust, but in infrared light, the galactic center reveals itself as a very complex place in which stars race around the radio source quickly. The closer the nearby stars are to the center, the faster they orbit it; one star, only a light week away from the source, showed visible motion in only two years of observation.

From these orbital characteristics, Andreas Eckart and his

colleagues have suggested that Sagittarius A* is a "supermassive" black hole with some 2.6 *million* times the mass of the Sun (Schwarzchild 1998). It seems that the peaceful, amorphous structure thought years ago to be our galaxy is no more.

Nebulae, and Star Birth in the Galaxy
Like most examples of the start of a life in Nature, the birth of a star is a secret event, hidden behind thick, dark clouds. The exciting story of the discovery of star birth within our galaxy began when the British astronomer William Herschel, some 200 years ago, first saw a dark nebula and exclaimed "My God! There's a hole in the sky!" By all appearances, Herschel was right: Since a dark nebula's thickness can block virtually all light from behind it, it looks like a window to the great beyond. Herschel's idea was that these objects might be "old regions that had sustained greater ravages of time" than their surrounding stars. His idea was good for its time, but it has since been improved upon. We now know that Herschel had stumbled onto a dark nebula, a mass of gas and dust in space, hiding stars behind it. In the mid-nineteenth century John Herschel, William's son, was cataloguing the southern stars from his observing site to add to his General Catalogue of Nebulae. What impressed him more than anything else were the dark regions he found in Ophiuchus and other parts of the Milky Way—areas so black, he noted, that they looked like gateways to something beyond. Deep in the southern sky he found an egg-shaped dark region that looks so black we now call it the Coal Sack.

Because there are no nearby stars to light these nebulae, they are dark. Edward Emerson Barnard, the American comet discoverer who was one of the most prolific observers ever, catalogued a large series of dark nebulae. He found them in two types of places in the sky: in the Milky Way, where we see them because they block out the light of more distant stars; and together with bright nebulae, where absorbing dust blocks the light of the brightly lit gas behind it.

Unless a dark nebula lies near a bright one, we cannot calculate how far away they are. If a nebula blocks out part of the stars from the more distant Milky Way, we know that it is closer than those stars. However, a dark nebula is more like a blind than an open window: its thickness blocks the view to the stars behind. It is dark because it is unlit by any nearby stars. A bright nebula is lit by stars that are near it. Probably the most famous example of a nebula appears in the sword of Orion. Through even the smallest telescope under a city sky, it is possible to detect the swirls of nebulosity that represent this nursery of stars. Through a telescope under a dark, rural sky, the nebula is stunning. Its brightest part lit by four bright blue stars in the center, the Orion nebula is a delicate mixture of bright and dark nebulae. It is truly one of the most magnificent offerings of the entire galaxy.

The sky is filled with these nurseries, where new stars are born. Many bright nebulae glow by emission, like neon lights: energy from nearby stars causes them to fluoresce. The Orion Nebula, the Lagoon Nebula, and the Eagle Nebula are examples of diffuse nebulae. Other nebulae do not fluoresce; because they consist primarily of dust, and not gas, they shine only by reflected light from nearby stars. The Merope Nebula, around one of the stars in the Pleiades, is an example of a reflection nebula. It cannot be seen without a telescope, but on a clear, dark night out in the country, a small telescope should reveal this cloud of the Pleiades.

Occasionally an older star will shed its outer atmosphere into space. As a result, a small nebula forms and surrounds a star. Because some of these clouds look like small distant planets, astronomers call them planetary nebulae. But they are not planets; they are temporary structures of gas. The most famous planetary nebula is the Ring Nebula, in the constellation of Lyra. Through a telescope, it resembles a smoke ring, a puff of gas released from a faint central star.

Very little is known about the biggest and widest dark nebula of them all. Straddling the Milky Way from Cygnus all the way to Sagittarius, it is called the Great Rift. Its length appears to divide the Milky Way into two parts. One of them trails off in the region around Ophiuchus, like an exit ramp to the Milky Way highway. We do not know much about this nebula, except that it is closer than the bulk of distant Milky Way stars whose light it blocks.

REFERENCES

Bok, B.J. and Bok, P.F. *The Milky Way* Cambridge: Harvard University Press, 1981, 170-172.

Burnham, Robert. *Burnham's Celestial Handbook* New York, Dover, 1966, 1978.

O'Meara, S. J. *Deep Sky Companions: The Messier Objects.* Cambridge: Cambridge U P, 2002.

"Spiral Arms of the Galaxy" *Sky and Telescope* 11, 6 (1952) 138-139.

Tinsley, B. M. "Theoretical Overview—Interactions Among the Galaxy's Components" Burton, W. B. *International Astronomical Union Symposium* No. 84: The Large-Scale Characteristics of the Galaxy (Boston: D. Reidel, 1979), 431.

Schwarzschild, Bertram. "Stellar Motion Very Near the Milky Way's Central Black Hole," *Physics Today* 51 (March 1998) 21.

Wordsworth, William. "I Wandered Lonely as a Cloud." *English Romantic Poetry and Prose* Ed. Russell Noyes. Oxford: Oxford U. P., 1956. 324-25.

COSMOLOGY 101

CAPTIONS FOR PAGES 129–132
For the captions to images on pages 133–136, see page 137.)

Page 129
Top: A Leonid meteor. California astrophotographer Wally Pacholka photographed this brilliant Leonid over Joshua Tree National Park.

Bottom: The launch of Apollo 8, December 1968. (NASA)

Page 130
Top: Earth from lunar orbit, taken from aboard the Apollo 8 spacecraft, Christmas 1968. (NASA)

Middle: Shortly after it was repaired in January of 1994, the Hubble Space Telescope took this image of all the fragments (except one) of Comet Shoemaker-Levy 9.

Bottom: Jupiter and Venus—the two brightest objects in the sky—at dusk. Fainter lights are stars. (Photo by David H. Levy)

Page 131
Globular cluster M15, about 40,000 light years away. Such clusters are usually found on the outskirts of galaxies, above and below the galactic plane. (NASA, STScI/AURA)

Page 132
Top: You never know what you'll get when you snap a shot of the diamond ring just before totality. Photographer Rick Jones explains: "This is the only picture I took of the (African) eclipse with my cheap little camera." The diamond part of the ring exploded into a cross-shaped beam of light in the camera. (Courtesy Rick Jones.)

Bottom: This Herbig-Haro object was formed when young stars ejected material into space. As these jets of material strike the nebulosity surrounding the new star at a speed of some 200 miles per second they produce shock waves that cause the gas to glow. (NASA, STScI/AURA)

COSMOLOGY 101

COSMOLOGY 101

COSMOLOGY 101

COSMOLOGY 101

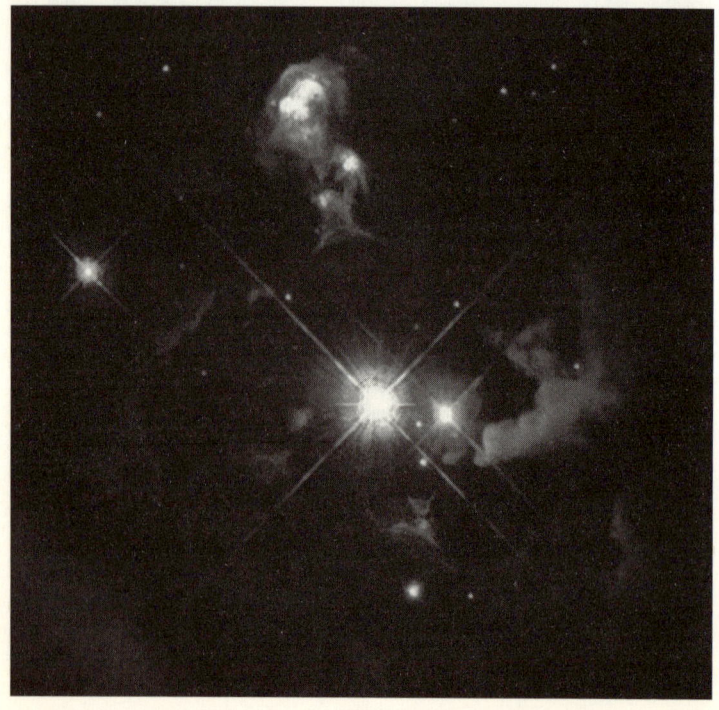

COSMOLOGY 101

COSMOLOGY 101

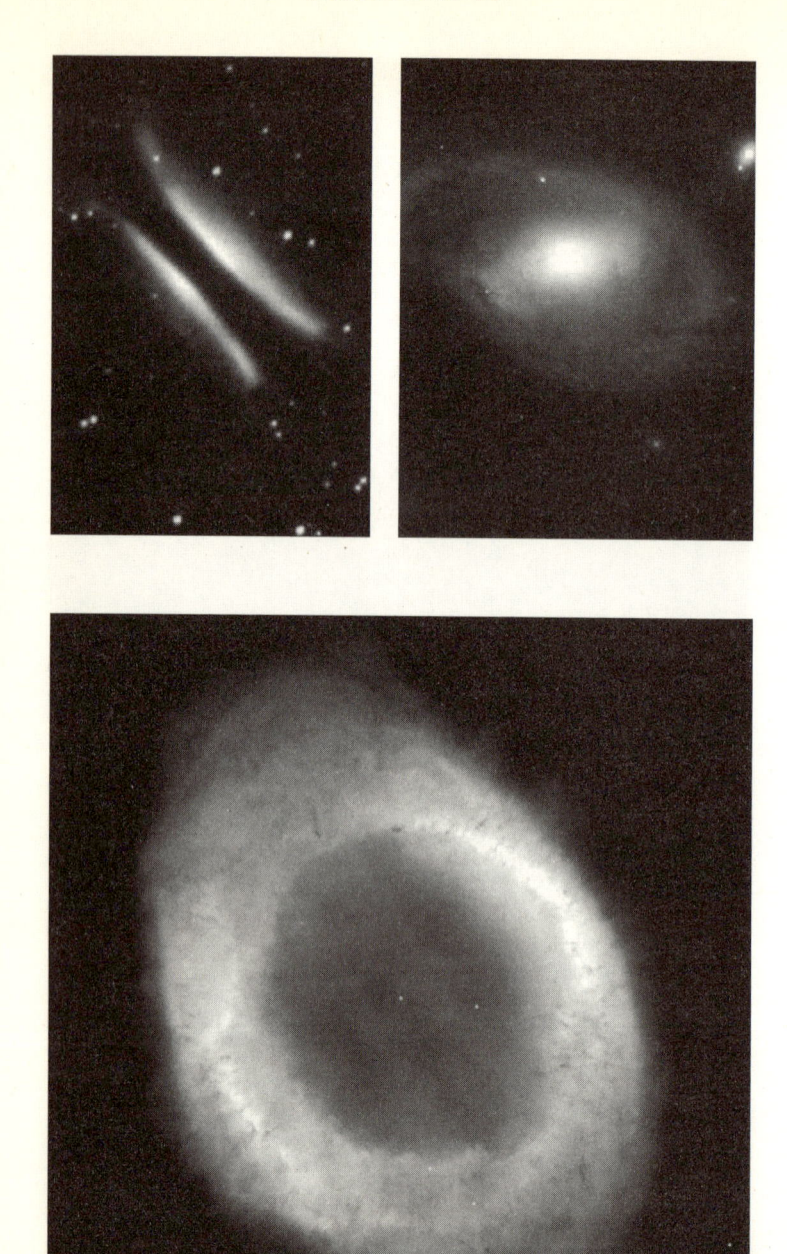

COSMOLOGY 101

COSMOLOGY 101

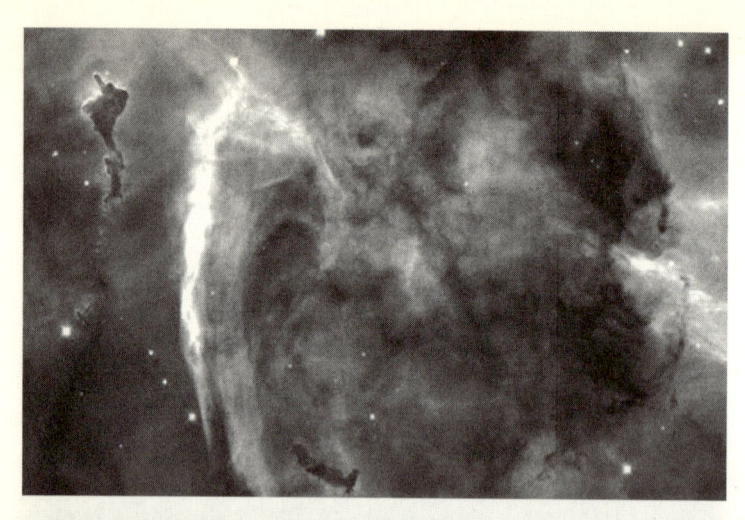

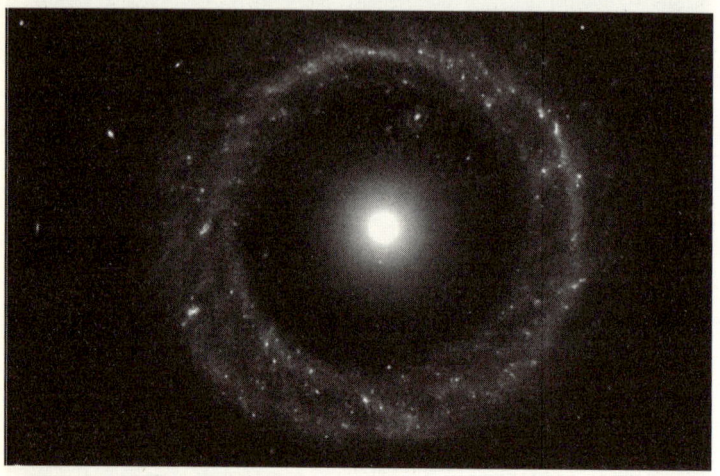

COSMOLOGY 101

CAPTIONS FOR PAGES 133-136

Page 133
Top: On June 26, 2001, the Hubble Space Telescope took this superb image of Mars at a distance of 43 million miles from Earth. (NASA, STScI/AURA)

Bottom: Taken in its natural color, the planet Saturn is stunning in this new image taken by the Hubble Space Telescope. (NASA, STScI/AURA)

Page 134
Top left: "Gomez's Hamburger" is a star in the stage immediately prior to becoming a planetary nebula. This image shows the dark band of dust, which is the shadow of a disk around the central star (unseen), edge-on. Named after its discoverer, astronomer Arturo Gomez. (NASA, STScI/AURA)

Top right: Although the galaxy at center, NGC 4319, and the quasar Markarian 205 (upper right) appear to be neighbors, they are not. NGC 4319 is 80 million light years away, but Markarian 205 is a *billion* light years from Earth. (NASA, STScI/AURA)

Bottom: The best optical photo ever taken of the Ring Nebula (also known as M57). It reveals details never seen before, including elongated dark clumps of material at the edge of the nebula and the central star floating in a hot blue gas. (NASA, STScI/AURA)

Page 135
The August 11, 1999 total eclipse of the Sun, photographed by Roy Bishop.

Page 136
Top: At the center of Eta Carinae is the Keyhole Nebula (pictured), an area of new star formation. (NASA, STScI/AURA)

Middle: Hoag's Object, discovered by astronomer Art Hoag in 1950. At first thought to be a planetary nebula, the objet is now known to be an unusual galaxy. The image shows a beautiful ring of blue stars, 120,000 light years wide, surrounding a galaxy. It is possible that the blue ring is the remnant of another galaxy that collided with the central one some two to three billion years ago. (NASA, STScI/AURA)

Bottom: Author David H. Levy at his own observatory with three different telescopes mounted and ready for the evening's observations. (Photograph by Wendee Wallach-Levy)

CHAPTER 11

Those Magnificent Galaxies

> . . . who at thy word bringest on the evening twilight, with wisdom openest the gates of the heavens, and with understanding changest times and variest the seasons, and arrangest the stars in their watches in the sky, according to thy will.
>
> —Ancient Jewish evening prayer, author unknown

The majesty of twilight—the peaceful setting of the Sun in the west while the dark shadow of the Earth rises in the east—is one of the truly remarkable effects of Nature. If the author of that prayer had only known just how many stars were arranged in the sky—all scattered through billions of distant galaxies and unimaginable distances—the amazement would have been quite beyond description. The writer would have marveled at the vast numbers of tiny spots of light beyond, each one a galaxy. The galaxies are grouped into clusters, and the clusters into vast assemblies called superclusters. In between, there is nothing but empty space.

THOSE MAGNIFICENT GALAXIES

The Realm of the Galaxies

Welcome to this realm, a place of unlimited dimension, where space is immense. The galaxies contain more stars than all the grains of sand on every beach on Earth. But from home, every star we see in the night sky belongs to our own Milky Way Galaxy. The *only* time in recorded history when this was not true was in the spring of 1987, when a star exploded in the Large Magellanic Cloud, the closest small galaxy to ours. The star became almost as bright as magnitude 2, the magnitude of the North Star—brilliantly visible in the southern sky yet shining from a distance of 169,000 light years.

In 1885, a supernova in the nearest large galaxy to us shone from the Andromeda Galaxy, some two million light years away. A sharp-eyed observer could barely see this star without a telescope. But unless such supernovae blaze forth, we cannot see any stars from other galaxies. I remember my dad once asking if all the stars in the constellation of Andromeda did not belong to the galaxy in Andromeda. I explained to him that Andromeda is a constellation like all the others, comprised of stars that are part of our own galaxy and generally not more than a few hundred light years away. However, within the boundaries in the sky of that constellation lies the very distant galaxy of Andromeda, whose stars are two million light years away. Thus, Andromeda is the name given to both a constellation and to a galaxy.

As we look out into space, we see that our galaxy is a member of a small clustering of 25 galaxies called the Local Group, and that our Local Group is, once again, at the edge of a mighty system called the Local Supercluster.

A Brief History

More than two centuries ago, the famous philosopher Immanuel Kant suggested that these distant objects were galaxies like the Milky Way. By 1800, William Herschel compiled a catalogue of some 2500 objects. Not all of Herschel's objects were galaxies, but most of them were. In 1850, two major events concerning galaxies occurred: Lord Rosse, an accomplished amateur astronomer and builder of a 72-inch telescope, then the largest on Earth, sketched the spiral shape of the galaxy we now call the Whirlpool, not far from the star at the end of the Big Dipper's handle. Also that year the German astronomer Humboldt suggested the name "island universe" for each of these distant objects. While today we follow Harlow Shapley's suggestion and call them galaxies, the idea of an island universe does have a romantic tinge to it.

In 1888, John Dreyer published the first edition of his *New General Catalogue* containing almost 8000 "nebulae" that optically appeared as fuzzy clouds, but virtually half of which were spiral galaxies. In 1908, two additional "index catalogues" appeared, raising the total number of catalogued objects to 13,000. Almost 90 percent of these fuzzy objects are distant galaxies. That number is paltry compared with what we have now seen, and what more modern catalogues include: galaxies in space virtually without number. And if our findings of small, irregular galaxies close to the Milky Way are any indication, a number of them are too small and distant to be seen.

The Great Debate
April 26, 1920, was a tremendous day in astronomy, a day focused on the National Academy of Sciences in Washington, where Harlow Shapley and Heber Curtis came together to discuss the state of the Universe. What were the spiral-shaped nebulae, and how far away were they? From his studies of the variable stars, Shapley knew that our galaxy was some ten times bigger than scientists had previously thought. From this, he thought that the spirals could not lie very far outside our galaxy. Curtis, using older reasoning,

thought that our galaxy was small, and that these remote spiral-shaped fuzzy patches were comparable in size and nature to our own galaxy.

An interesting debate indeed: Although Shapley's reasoning about the size of our galaxy was correct, he was wrong about the nature and distance of the spiral nebulae. Curtis was right about the spirals, but for the wrong reason! Using the observational data of the time, both scientists did the best they could. With the available telescopes, Shapley could peer no farther than the globular clusters, which are really at the outskirts of our own galaxy. Just four years later, Edwin Hubble used Shapley's own yardstick, the variable stars, to prove Shapley wrong in the great debate. Using the newly opened 100-inch telescope on Mount Wilson, Hubble found stars within these spirals. These stars were so far away that the spiral nebulae in which they lived had to be distant galaxies. Today we use the Hubble Space Telescope to find these variable stars in those galaxies. In this way, the work Shapley did three-quarters of a century ago continues on. As long as telescopes get bigger and better, we will be able to measure distances farther and farther out.

How Galaxies Came to Be
The early Universe was a soup consisting of hot gas and unseen matter. Although this soup was quite homogeneous in structure, some areas were denser than others. The dense regions had slightly more gravity than the regions surrounding them, so that eventually these areas began to circulate, pulling in the matter around them to form massive protogalaxies. Matter within these protogalaxies began to contract toward their centers, and spin more quickly, just like a figure skater spins faster as she brings her arms closer to her body. After most of the gas settled into a galaxy's center and surrounding disk, it began to condense further to begin the process of star formation. The gas was spinning so rapidly

that it could not fall to the center; consequently, much of it turned into stars in the galaxy's disk and spiral arms. The faster the gas was spinning, the larger in area the resulting galaxy became. A protogalaxy spinning rapidly forms a galaxy of larger area than one with the same amount of gas spinning more slowly. Within the relatively brief time scale of perhaps a few hundreds of millions of years, galaxies began to form in great numbers.

Our Local Group
How bright do the galaxies really look from the comfort of our own observing places on Earth? In the constellation of Andromeda is the brightest galaxy yet discovered. This galaxy is known as Messier 31, or the Andromeda Galaxy. The Milky Way and Andromeda are the two largest galaxies of a 25-member-small clustering called the Local Group. A third galaxy, Messier 33, lies in the constellation of Triangulum. All the other galaxies in the group are small ones. Two of these are so close to the Milky Way—about 160,000 light years—that they can readily be called satellites of it. They are the Large and Small Magellanic Clouds. As we have already seen, the Large Magellanic Cloud has provided scientists with a good deal of information about stars, for in 1987 the brightest supernova in almost 400 years appeared there.

Our Local Group belongs to a throng of clusters of galaxies collectively know as the Local Supercluster. The member clusters are in the constellations of Ursa Major, Coma Berenices, Leo and Virgo. If you point a small telescope into the center of the Virgo cluster and then move it about slowly, you might see several galaxies.

Messier's Catalogue
Scattered throughout the sky are 110 "deep sky objects" that were collected into a catalogue by a famous eighteenth-century French observer, Charles Messier. Some of them are

nearby galaxies, and others are clusters of stars or clouds of gas and dust.

Messier wanted his claim to fame to be the comets he discovered. The first person to find comets as part of an organized telescopic search, from 1760 to 1770 Messier was virtually the only discoverer around. He was so prolific that Louis XV of France called him the Comet Ferret. Messier's good fortune ended with the storming of the Bastille and the start of the French Revolution in 1789. Although he continued his searching, he was without the pension that allowed him to live while he spent his time hunting for comets. Virtually penniless, he even had to borrow oil for his observing lamp from Lalande, one of his friends. During his years of searching Messier kept a careful listing of the fuzzy objects—nebulae, star clusters, and galaxies—to avoid confusing them with comets. Messier's comets are long gone, but the members of his catalog are forever present in the sky. Of the galaxies listed below, almost all are designated by "M," or Messier, numbers. Each time you look at one, you will be seeing the same object that gave Messier an initial thrill at the thought of possibly finding a new comet. In the sky, comets and galaxies have similar appearances. The fact that the two types of objects are so different, despite their fuzzy similarity, is one of the marvels of astronomy: a comet appears as a cloud of gas and dust the size of a planet, while a galaxy is a swarm of stars hundreds of thousands of light years wide and millions more away. But both types of objects have a fuzzy appearance.

Types of Galaxies

Considering the vast numbers of galaxies, it is somewhat surprising that there are only a few different types. The giant *spiral* systems are the most common type; another is *elliptical*. Small galaxies tend toward random shapes; we call these galaxies *irregular* in type.

COSMOLOGY 101

The spiral galaxies are among the most beautiful in the sky. They are also the most common: some four-fifths of the galaxies we know of are spiral in form. Their graceful spiral arms stretching out dramatically, these galaxies offer many hours of pleasant viewing through a telescope. According to Edwin Hubble's classification, spirals range from being very tightly wound (Sa), to moderately wound (Sb) (which is what we used to believe our galaxy to be), to very loose, with arms spread over a wide area (Sc). Some galaxies have a central bar in addition to a fattened nucleus, with the arms spreading out from the ends of the bar. New evidence suggests that the center of the Milky Way might actually be a short bar, and that our galaxy is a barred spiral.

Some Spiral Galaxies

M31. The Great Galaxy in Andromeda. By far the most beautiful galaxy in the sky. On a clear dark night away from city lights M31 can be seen with the naked eye; it is in fact the farthest object usually visible without a telescope.

M33. The Pinwheel Galaxy. This is a face-on spiral galaxy in Triangulum.

M51. The Whirlpool Galaxy. Galaxies present themselves in a number of aspects, face-on, edge-on, or somewhere in between. Found near the end star of the Big Dipper's handle, the Whirlpool Galaxy is a fine example of a galaxy seen face-on. It looks double because a second galaxy appears to hang on from one of the arms.

M58. A bright barred spiral galaxy in the constellation of Virgo.

M63. The Sunflower Galaxy. This is a bright, elongated spiral in Canes Venatici.

M64. The famous Black Eye Galaxy in Coma Berenices. Through moderate-sized telescopes on a good night, this galaxy boasts what seems to be a black eye near its center.

Actually the black represents a large area of dust unlit by any stars.

M65 and **M66**. Two spiral galaxies close together, near the tail end of Leo.

M81. With M82, also known as Bode's Galaxies. Large, striking spiral galaxy in Ursa Major.

M83. A beautiful spiral galaxy in Virgo, near the constellation of Corvus.

M88. Richly beautiful galaxy in Virgo.

M101. The Pinwheel spiral Galaxy in Ursa Major.

M104. The Sombrero Galaxy in Virgo, a beautiful galaxy that actually bears a resemblance to the Mexican hat.

Elliptical Galaxies

Far mightier than the spirals, ellipticals can be overwhelming in size. Hubble classified the ellipticals from E0 (perfectly round) to E7 (very oval). Less than a fifth of the known galaxies are elliptical.

M87. This galaxy in Virgo contains some five trillion stars, making it one of the largest in the Universe. In its center might be a black hole a *billion* times the mass of the Sun, with a strong jet of gases shooting out of it.

M84 and **86**. Two fainter ellipticals in the midst of the richest and most beautiful region of galaxies in the sky, in Virgo.

M49. A bright elliptical galaxy in Virgo.

M60. Bright elliptical, also in Virgo.

Irregular Galaxies

Too small to evolve into an organized spiral or elliptical shape, the irregulars used to be considered the least populated class, with fewer than 3 percent of the known galaxies being classified that way. However, recent discoveries of dwarf irregular galaxies in the constellations of Antlia and

Carina seem to indicate that the Universe may be filled with many times more of this type of galaxy than any other type. They are so small than we can see only the closest samples. Our own Local Group contains only three large galaxies, but they are accompanied by more than 20 smaller, irregular ones.

Close to the bright spiral M81, M82 has several dusty areas blocking our view. From much of the Northern hemisphere, Messier 82 is visible most nights of the year. This galaxy, tortured with the spread of gas involved with the birth of new stars, is the most unusual of the bright galaxies that we can see with a small telescope. Whenever I look at it through the highest power of my 16-inch diameter reflector, it dominates my field of view with a sense of otherworldliness and mystique. What strange processes must be going on there! But among the trillions of galaxies the Universe must offer, a Universe of expanding superclusters of galaxies seemingly without end, this one should hardly be the most unusual.

The Small Magellanic Cloud. At 72 degrees south declination, this small companion to the Milky Way is visible only from southern locations.

NGC4214. A large irregular galaxy in Canes Venatici.

NGC4449. Also in Canes Venatici, this galaxy has a rectangular shape.

How Galaxies are Distributed in Space

With the opening of the Mount Wilson 100-inch telescope in 1918, astronomers had the most fantastic tool in the history of astronomy to study the distant expanses of the Universe. Edwin Hubble took advantage of this telescope to survey random areas of the sky to see how distant galaxies were distributed. Although his telescope was large (its light gathering mirror more than eight feet wide), its field of vision was narrow. Hubble sampled small areas around the sky and concluded that the clusters of galaxies were distributed evenly through space. And there was what he called a "zone of avoidance" where the Milky Way blocks any view of a galaxy located behind it.

An astronomer with little knowledge of galaxies, but with a good understanding of the sky, came to a different conclusion. During his long search for distant planets in the 1930s, Clyde Tombaugh of Lowell Observatory took notes on the distribution of galaxies in each of his photographs. Instead of sampling the sky, Tombaugh photographed all of it. His conclusion: Galaxy clusters are not distributed evenly at all, but are clumped together.

In the mid 1940s, Tombaugh and Hubble debated the issue in Hubble's office at the California Institute of Technology. Tombaugh explained how, on his photographic plates covering all of the sky visible from Flagstaff, he did not see an even distribution of galaxy clusters. Tombaugh reported dense concentrations of galaxies on his films, especially a mighty "stratum" of galaxies stretching from the constellation of Pegasus, through Andromeda, and into Perseus. He also reported "voids," where he saw little but empty space where there should have been galaxies. The discussion left no winner; Hubble simply didn't take the younger scientist seriously. He should have. A decade later, George Abell of Palomar Observatory, in a doctoral thesis using photographic plates from the new Schmidt camera at Palomar Observatory, confirmed Tombaugh's view of the Universe, and went on to show that the clusters of galaxies are clumped into vast superclusters. It appears that at the supercluster level, the distribution is indeed uniform. The number of stars in all these galaxies is virtually without end, probably greater than the number of grains of sand on all the beaches of Earth.

Farther and Farther Galaxies

In March of 1998, a team led by Arjun Day, using the 10-meter diameter Keck II telescope on the Hawaiian island of

Mauna Kea, announced the discovery of the farthest galaxy seen to that time. Listed as 0140+326RD1, (a symbol for its position of 1 hour 40 minutes, and 32.6 degrees north declination) this tiny, barely visible galaxy in the constellation of Triangulum may be fewer than a billion years younger than the Universe itself. That means that we are seeing this galaxy as it appeared several billion years before our Sun was born. Probably none of its stars still exist, at least not as they appear to us now. What civilizations flourished in that maelstrom of stars so long ago? And as their scientists looked out on a Universe far smaller than ours, did they wonder as we do, how it all began and if it will ever end?

Galaxies farther and farther afield are discovered, it seems, virtually with each passing month. Soon we might find one so far back that it is barely complete, a protogalaxy still in the process of condensing into its first generation of stars. When we see such a galaxy, we will really be looking at our own creation.

CHAPTER 12

Galactic Exotica

Tiger! Tiger! burning bright
In the forests of the night,
What immortal hand or eye
Could frame thy fearful symmetry?

In what distant deeps or skies
Burnt the fire of thine eyes?
On what wings dare he aspire?
What the hand dare seize the fire?

And what shoulder, and what art,
Could twist the sinews of thy heart?
And when thy heart began to beat,
What dread hand? and what dread feet?

What the hammer? what the chain?
In what furnace was thy brain?
What the anvil? what dread grasp
Dare its deadly terrors clasp?

When the stars threw down their spears,
And watered heaven with their tears,
Did he smile his work to see?
Did he who made the Lamb make thee?

Tiger! Tiger! burning bright
In the forests of the night,
What immortal hand or eye
Dare frame thy fearful symmetry?
                              —William Blake

F ar off in the depths of space and time lie great cauldrons of light and darkness where the laws of gravity conspire to create a zoo of specimens that boggle the mind, an unbelievable collection that of zebras, giraffes, and tigers in the night. These objects are not the kinds of things that we see around us, but they are real, and they are what's out there. Previously, we explored various types of "normal" galaxies. Now we'll look at some of the stranger types.

Strange Spirals
While spiral galaxies typically have a bright center and two or three long arms, there are some wild exceptions. NGC 613 is a faint spiral in the constellation of Sculptor, deep in the southern sky. Although it displays the beautiful symmetry that is true of most spirals, this galaxy has at least five arms spreading off into space (Arp and Madore, 1987).

NGC 7332 in Pegasus is a lenticular (lens-shaped) spiral galaxy bereft of arms altogether. Although it has a central bulge and disk, there is no evidence of spiral structure, and no evidence of any population I material like hot O and B stars, young clusters, and gas and dust. Astronomers first thought that these galaxies did have arms at an earlier stage of their evolution. Without any new raw material for star formation, these galaxies seemed to be close to their final phase of life. However, new evidence shows that they do contain some raw material for future star formation. NGC 2685 is one of the most unusual galaxies. Surrounding its long cigar-shaped body are no spiral arms; instead there are a series of helical filaments that almost look like puffs of smoke at right angles to the long "cigar" axis.

Very Large Elliptical Galaxies
In 1964 William Morgan (who first determined our own galaxy's spiral shape) described a rare and different kind of

elliptical galaxy that tends to be found near the centers of clusters of galaxies. Very large and luminous, these A-D galaxies are surrounded by huge and very diffuse envelopes of stars. Some D galaxies even have more than one core or nucleus. Their name is part of a classification by Morgan who used the designation A-D to include ellipticals that are unusually large and (D) diffuse. The prefix refers to the largest examples.

Markarian Galaxies
Literally one of the Universe's hottest subjects, these bluish galaxies were first catalogued by the astronomer Benjamin Markarian. These galaxies are identified not by visual observing but through the excess blue light, and strong ultraviolet emission, that shows up in their spectra. With 1500 such galaxies found so far, this type is not that rare, and members of the group exhibit different types of activity in their cores.

Starburst Galaxies
In galaxies like Messier 82, star formation is taking place on a grand scale at the galaxy's center. These galaxies are unusual for a simple reason: large numbers of their stars happen to be forming at the same time. A subset of this type is the "dwarf compact galaxy," which is a small irregular galaxy where bursts of star formation dominate the entire system. These dwarfs are also known by the term "isolated extragalactic H II region." These galaxies have bright regions of ionized hydrogen gas out of which stars are born.

There is one isolated extragalactic H II region in the constellation of Leo. It is an enormous region of neutral hydrogen that lies in a dense region of galaxies near Messier 105 (Schorn 1988). If it were near our galaxy, it would stretch clear across to the Andromeda galaxy, some two million light years away.

Radio Galaxies
Some galaxies are very bright when recorded with radio telescopes, even though they are no brighter than ordinary galaxies in visible light. These "radio galaxies" are usually giant elliptical systems that typically have a protrusion of light emanating from their centers. The most powerful such galaxy is called Cygnus A, an object earlier thought to be two colliding galaxies. These galaxies might indeed have been colliding at one time, but current models indicate that Cygnus A is an elliptical galaxy with a dust lane that is the remnant of an episode during which it cannibalized a smaller galaxy long ago.

Radio galaxies often appear double-lobed. These galaxies emit large amounts of radio energy from two lobes on either side of the galaxy. The lobes appear to be explosive releases of energy that fly off along the axis of the galaxy's rotation. They are very much larger than the galaxies they surround; in fact it is not unusual for the distance from one lobe to the other lobe to cover more than 200,000 light years, which is twice the size of our galaxy. The largest one spans some 19 million light years!

Centaurus A is a choice example of a double-lobed radio galaxy. At its center is a bright elliptical galaxy called NGC 5128. Through a telescope, this galaxy seems to be divided by a large area of dust that is a region in which star formation is in progress on a massive scale. Because the galaxy is relatively close to us, the surrounding lobes cover a huge area of sky, equivalent to the bowl of the Big Dipper!

Colliding and Interacting Galaxies
The effect of two galaxies colliding with one another must be astonishing. Although it would seem that the stars in a galaxy are so far apart that they would just flow past each other, their hydrogen gas and interstellar dust reacts in a collision. About a dozen "ring galaxies" have been photographed.

GALACTIC EXOTICA

Arp 321—aka Moe, Larry and Curly. I found this cluster of galaxies—they numbered three in my telescope—while comet-hunting one evening.

Strange systems consisting of a central core surrounded by a ring, these galaxies are thought to be the result of a galactic collision in which the smaller one is totally disrupted to become the ring around the larger one.

Two of the most exciting colliding galaxies are NGC 4038 and NGC 4039 in Corvus, together called the Antennae. If our understanding is correct, about a billion years ago these galaxies were several hundred thousand light years apart. Their outer edges began interacting about 500 million years ago. As they closed in on each other, tails of matter were driven out from the galactic centers by intense tidal forces. The two tails resemble a curved rabbit-ear TV antenna.

When it comes to colliding galaxies, we cannot ignore our own Milky Way! It has been known for some time that the Milky Way is interacting gravitationally with its two neighboring galaxies, the Large and Small Magellanic Clouds, and recent models indicate that the Milky Way could cannibalize the Large Magellanic Cloud some time in the future. Though they are well known, the clouds are by no means the only nearby galaxies. One recent discovery is a dwarf spheroidal galaxy in Sagittarius. This galaxy is hard to see because it lies on the other side of the Milky Way's center. Although it is only a tenth the diameter of the Milky Way, it has been orbiting our galaxy from a distance as close to the Milky Way's center as we are! As it orbits the center, once every billion years, it forces its way through some of the Milky Way's densest regions. Why doesn't this galaxy just fall apart, releasing its stars to becoming part of the Milky Way? It has survived this long because it contains a large amount of dark matter, the combined gravity of which strengthens the system and hold its stars together. The galaxy has probably survived at least ten orbits, but it cannot win its engagement with the Milky Way. Eventually our galaxy will cannibalize it (AAS release, 1998).

Active Galaxies
We usually think of galaxies as massive and distant things, and we do not expect them to change their appearances quickly. So when, in the 1950s, the astronomer Carl Seyfert noticed that the bright cores of some galaxies actually did change in brightness slightly over months or years, they were astonished. Seyfert, who was killed in a car crash in 1960, never lived to appreciate the importance of these galaxies,

and never knew that this important class of galaxies is now named after him. The Seyfert galaxies, of which Messier 77 is the best known example, are great spiral systems with extremely bright centers. Their cores actually vary in brightness in a leisurely fashion over a multi-year cycle.

Seyfert galaxies are thought to be an intermediate form of galaxy between normal galaxies like ours and the great quasars. Another intermediate form is called the BL Lacertae objects. BL Lacertae was originally thought to be a slowly varying star within our own galaxy. We now recognize that it is the very bright center of a distant elliptical galaxy. Although these cores must contain hundreds of millions of solar masses, they can vary in brightness by a factor of five in a single day!

Hell's Heart

The quasars are the true celestial tigers of the Universe. Short for *quasi-stellar radio source*, a quasar is an extreme version of a Seyfert galaxy—the highly energetic core of an extremely distant spiral galaxy. While the cores of Seyfert galaxies are several times brighter than the surrounding disk and arms of the galaxy, quasars are thousands of times brighter than the rest of the galaxies they inhabit. If the quasars we see were ordinary galaxies, like Andromeda, we wouldn't detect them at all because of their great distances. But if our own Milky Way had a quasar at its center, we would see it blaze as brightly as the full Moon. Even at their great distances, measured in billions of light years, some quasars appear as faint stars when seen through large home telescopes.

Discovery of the Quasars

The year 1963 was an extraordinary one: President John Fitzgerald Kennedy was murdered while waving at crowds in downtown Dallas, the Beatles were poised to lead the

British rock invasion of America, and at the edge of the Universe, the faint light of the most monstrous objects ever detected was being received, and finally understood, on Earth.

Had astronomers not had access to two different types of telescopes, optical and radio, the true nature of quasars might still elude us. In 1960, radio telescopes had revealed a strange source called 3C 48, that seemed to coincide with a faint blue star. It was the first "star" (other than the Sun) that seemed to emit radio waves. Some time later, another bright radio source, 3C 273, was occulted by the Moon three times within a period of several months. As the Moon passed in front of the radio source, the far-off object's "noise" was cut off abruptly. Because we know the position of the Moon precisely at a given time, these events allowed astronomers to determine the exact position of the radio source. In 1963, Maarten Schmidt turned the largest telescope in the world, the 200-inch device on Palomar Mountain, to the position that had been derived by the lunar occultations. He found a fairly bright star that was accompanied by a fainter jet of light. Not quite stars, these objects were officially designated as quasi-stellar radio sources. (This was later shortened to *quasars*.) In 1963, Schmidt obtained a series of spectra of this object, and realized that if its redshift were interpreted correctly, that 13th magnitude star with jet was billions of light years away, and was receding from us at almost a fifth of the speed of light! Soon after, Jesse Greenstein suggested that the redshift of 3C 48 (Object 48 in the Third Combined Catalogue of Radio Sources) meant that it was rushing away from us at more than twice that velocity.

As the world mourned the death of the American president, the news services crackled with the startling words that astronomers had found a new kind of object that stretched the limits of the Universe farther than anyone could have imagined. To a saddened world, the discovery of the quasi-stellar radio sources provided excitement and hope from afar.

About Quasars

One of the brightest quasars, called 3C-273, is so bright that it can be spotted through a small telescope—if you know just where to look and, with the aid of an atlas or computer program, can distinguish it from the surrounding stars. Large telescopes reveal much more detail, including the existence of a jet of material racing out of the quasar's center.

What would it be like to sail past this tiger in the night? We physically could not get very close to it, for its energy is so strong that it would destroy us even if we were a few hundred light years away. But we can imagine an energy machine so vast that it can create black holes simply from the collapse of dust clouds. A quasar's heart is not very much larger than our solar system. The more matter it has, the more energy it releases. It is a powerful energy source like a supermassive black hole, which swiftly pushes electrons up to high speeds as they spiral outward. The electrons emit what is call synchrotron radiation. Surrounding it are rich clouds of gas that have erupted out of the center in times past.

Gravitational Lenses

In late 1979, two University of Arizona professors tried a fun experiment while having drinks. They carefully broke their wine glasses, but saved their bases with stems attached, and conducted an experiment—one that you can try as well, if you're willing to sacrifice a wine glass. Holding the stem so that the base is away from you, point it toward a light source. Watch how the base acts as a lens. Depending on how you hold the stem, the base will divide the light into two or more separate lights. Or if you hold the base exactly right, there will be a ring of light around the base.

This simple demonstration illustrates one of the most exciting discoveries of modern astronomy. Found in the late 1970s on a Palomar Sky Survey photograph was a pair of identical quasars. Then, with the power of the then-new Multiple Mirror Telescope south of Tucson, astronomers found that the quasars' redshifts were virtually identical. Could both images be of the same object? And if so, what was acting like a broken wine glass out there in space that could be bending the light?

When astronomers attached an electronic CCD (charge-coupled device) to the telescope, they found their lens. Directly between the distant quasar and us lies an extremely faint galaxy. Ecstatic astronomers had their answer: The galaxy and the quasar were playing out an important part of Einstein's general theory of relativity—that gravitational sources could bend light. The gravity of the galaxy was acting as a lens, bending the light of a more distant quasar so that it appeared double.

Since then astronomers have identified even more exotic examples of gravitational lenses. Some galaxies are so perfectly placed along our line of sight that they cause more distant objects to break up into beautiful arcs and even rings. And one galaxy is so elegantly positioned that the quasar so far behind it is broken up into four images we now call Einstein's Cross.

What incredible specimens we have observed in this universal zoo! These objects are so far off that we know little about them. Forcing us to learn about them by analyzing their faint specks of light, our Universe reveals its secrets slowly and patiently. From the massive star formation going on in a starburst galaxy to the sight of two galaxies in collision, the Universe behaves as one dysfunctional family. And as we culminate with a quasar, the Universe's Tiger burning bright, we see that wondrous things must be happening in the fearful symmetry of its heart of hell.

GALACTIC EXOTICA

REFERENCES

American Astronomical Society. "Invader Galaxy Apparently Contains Much Dark Matter." AAS news release 03 Feb. 1998.

Arp, Halton, and Barry Madore. *A Catalogue of Southern Peculiar Galaxies and Associations Cam.* Cambridge: Cambridge U. P., 1987.

Blake, William. "The Tiger." 1794. *The Norton Anthology of English Literature.* Ed. M. H. Abrams et al.. Vol. 2. New York: W.W. Norton, 1962. 55-56.

Schorn, Ronald. "The Extragalactic Zoo." *Sky & Telescope* 75 (1988): 27.

CHAPTER 13

A Modern Cosmology

> Goe, and catche a falling starre,
> Get with child a mandrake roote,
> Tell me, where all past yeares are,
> Or who cleft the Devil's foot.
> Teach me to heare Mermaides singing,
> Or to keep off envies stinging,
> And find
> What winde
> Serves to advance an honest minde.
> —John Donne

These beautiful childhood dreams and wishes, so appropriate and wistful today as they were almost four centuries ago, offer a beautiful way of inquiring about cosmology. How does the Universe work? Where are we in it? How did it all begin, and where is it headed?

We live on a *planet* that orbits a medium-sized *star* about two-thirds of the way out on one of the *spiral arms* of a *galaxy*, which is part of small *local group* of more than a dozen other galaxies, which in turn is part of a *Local Supercluster* of galaxies. By turning to the galaxies in Coma Berenices and Virgo, we see other components of this *local supercluster*. Much further out, the Universe is a tapestry of other superclusters that keep going on and on—all racing away from us at incredible velocities.

All this makes us seem awfully small, doesn't it? It shouldn't. To think that we are a part of all this action is a thrilling thought, even more so when we imagine that all parts of the Universe, big or small, are somehow integral to its operation. The idea that we have the intelligence to pose questions about it is the best of all. This ability is a wonderful part of the Universe; the human brain may be the most complex item we have so far found in this Universe, more complex even than a star or a galaxy that works using, essentially, gravity. Are there clusters of superclusters, and clusters of clusters of superclusters? And what is the future of the superclusters? Will they expand forever? And how did it all start?

If the Big Bang model of cosmology is correct, in the very beginning of the time and space we know, all the matter and energy of the Universe, as well as the space within which it would later expand, was contained as a single point. What was it like in this miraculous point, which contained everything? That single point could not have lasted more than a nanosecond, and the explosion that followed marked the birth of our Universe. It was not a point inside space, but an eruption that created space. Today, our Universe continues to expand like a loaf of raisin bread: As the loaf rises, the raisins, each one representing a supercluster of galaxies, appear to move away from each other. But the superclusters did not exist in the early Universe.

After the explosion, our Universe was a soup of elementary particles and gamma rays, rapidly expanding and cooling at the same time. It was composed of both matter and antimatter—a kind of matter in which atomic particles are opposite in charge and other properties, from those in matter. However, since matter was stronger in what we call trace amounts, it won out, and so our universe became a matter universe, and not an antimatter one.

Inflation?

Is it possible that, although the Universe began as a point, its scale inflated like a balloon during its first tiniest fraction of a second of life? This "Inflationary Universe" theory suggests that the Universe expanded enormously right after the initial Big Bang, and then cooled catastrophically in a process called supercooling. We see supercooling often when we look up at a cloud. If the air has been cooling rapidly, the cloud might not turn from water vapor to ice crystals as soon as the temperature drops below the level at which water freezes. The temperature drops so fast that ice crystals do not have a chance to form right away.

The Inflationary Universe theory posits that the Universe grew so fast that it supercooled as it inflated. The result: Just as supercooled clouds have excess energy in the form of liquid water that really should be ice crystals, the whole Universe had excess energy, which quickly changed to matter and electromagnetic radiation. Thus, the Universe was given an enormous energy gift by this simple process.

Quarks and Universal Forces

As any high school student knows, atoms comprise protons, neutrons, and electrons. What high schoolers might not know, however, is that electrons in turn are made of *leptons*, and protons and neutrons, in turn, are made of *quarks*. The name was appropriated from James Joyce's novel *Finnegan's Wake*. A subatomic particle, quarks are divided into six types, called (really!) *up*, *down*, *strange*, *charmed*, *truth*, and *beauty*. Quarks are held together by a stringy glue called (naturally) *gluon*. The *nuclear force* (or the strong force) involves the quarks and gluons. At some early stage of the Universe, this force was dominant. Gravity is much weaker than the nuclear force; however, the force holding an atomic nucleus together diminishes to nothing long before it reaches the neighboring nucleus. Gravitation diminishes

A MODERN COSMOLOGY

much more slowly, so on the scale of the current Universe it is gravity that determines the structure.

The First Second

Quarks, antiquarks, and photons occupied the Universe just before its first thousandth of a second. As the primordial soup thickened, the Universe still expanded, allowing quarks to combine into pairs and triplets, and then into protons and neutrons. Quark and antiquark—matter and antimatter—existed in almost equal numbers at this time. The "war" between them was won by a small surplus of quarks by the end of the first second, and the quarks have begun to gather to form protons and neutrons. Another basic power, the *electromagnetic force*, came into play at this time. Like *gravity*, it also diminishes slowly, but it was and is much stronger than gravity. However, because protons and electrons have precisely equal and opposite charges, the electromagnetic force is normally canceled out. If the charges were not equal, the electromagnetic force, and not gravity, would have determined the large-scale structure of the Universe. There would have been no galaxies, no stars, and no planets, just unformed energy and lightning.

A Few Minutes Later

Although the Universe was cooling rapidly, it was still as hot all over as the center of a star. During these minutes, protons and neutrons fused together to form "heavy hydrogen" (deuterium) and alpha particles, or helium nuclei. But with a continuation of the cooling process, within a few minutes at most, fusion shut down.

Three Hundred Thousand Years Later

From the end of the first three minutes to the end of the Universe's first 300,000 years of existence, cooling was the order of the day. By that time the temperature of the entire

Universe had dropped to less than 3000 degrees F, it became transparent to photons of electromagnetic energy, and they were set free. We can detect these primeval photons as something called background radiation. It appears that this background radiation, which we will discuss later and which is visible at radio wavelengths all over the sky, is the earliest indication we will have of the early Universe, for it represents the time that it first became detectable. But what happened in the next half billion years? As mentioned earlier, a galaxy has been found that appears to date from a half billion years after the Big Bang. The most distant quasars seem to be only about a billion years younger than the Big Bang. As their light passes through invisible clouds of gas that lie between them and us, it changes its spectral appearance. The spectral lines of most of the distant quasars show evidence of vast clouds of hydrogen. Graphs of quasar spectra show forests of small "Lyman-alpha lines" that denote the presence of hydrogen atoms in masses of cosmic clouds lying at various distances en route to the quasars. First observed by Roger Lynds at Kitt Peak National Observatory in 1971, these lines might be the telltale signatures of gas fragments that were beginning to form galaxies in the first billion years of the Universe. These clouds appear to be very large—by observing similar clouds as seen through the light path of two closer quasars, astronomers estimate that they could exceed 350,000 light years in diameter, far larger than a typical galaxy. Although some of the closer clouds might be associated with galaxies, others seem to hang alone in the depths of space.

If these Lyman-alpha clouds are the precursors of galaxies, then we have spotted an important marker for the Universe from before the time of galaxies. However, there is still some question about whether these galaxies were formed out of the clouds or the clouds were created by the galaxies.

A MODERN COSMOLOGY

With better detectors and telescopes, in the near future we might have some answers.

The Background Radiation

Using the data from a simple antenna three decades ago, and a small satellite in orbit, cosmologists have outlined the shape of the Universe in space and in time. The existence of a small amount of background radiation, at 2.7 degrees above zero and existing all over the sky, had been predicted as early as the late 1940s. Because the Big Bang occurred throughout the Universe at its birth, and the Universe has been expanding ever since, the remnant radiation should be detectable everywhere in the Universe. This radiation was discovered quite by accident. In 1965, Arno Penzias and Robert Wilson of Bell Telephone Labs were preparing a small radio telescope to receive communications from artificial satellites. They were troubled by some unexplained static in the 7-centimeter band. The static seemed to be coming from all over the sky, all the time. Even after they tightened the telescope's connections, checked for loose wires, and even cleared away detritus from nesting pigeons, they could not get rid of the faint static. When they eventually isolated the cause, they were amazed. Their static was coming not from pigeons a few yards away, but from ten to twenty billion light years away! The radiation was consistent with a source radiating at only three degrees above absolute zero. That unknown source turned out to be the primordial background radiation, coming from the entire sky. They had caught the very echo of the colossal explosion that began the Universe.

Penzias's and Wilson's astonishing discovery earned them the 1978 Nobel Prize in physics. One scientist reminisced how, when major discoveries are made, we are reminded that Einstein believed that our situation on Earth is unique.

"Everyone of us appears involuntarily and uninvited for a short stay without knowing why," the great scientist mused, and then concluded, "To me, it is enough to wonder at the secrets."

In April 1992, NASA's Cosmic Background Explorer (COBE) satellite was studying this background radiation from space. The satellite's instruments detected small variations in the temperature of the background radiation. These variations are the seeds of future galaxies, deposited just 300,000 years after the Big Bang but still lying round space for us to detect. These fluctuations within the background radiation were incredibly small, in the range of 3 hundredths of a degree—barely detected, but they were later confirmed by other types of observations.

The Cosmological Principle and The Hubble Parameter

If the Universe is expanding from a uniform point, it should look about the same as viewed from anywhere in the cosmos. This theory, conceived by Edwin Hubble, is known as the *cosmological principle*. The principle works only on extremely large scales, of several hundred million light years or larger. In other words, nearby galaxies are not distributed evenly, nor are clusters of galaxies. But Hubble believed that the *clusters* of galaxies were distributed evenly throughout the Universe. This is true only at the supercluster level, and recent discoveries of a Great Wall and a Great Attractor indicate, as we shall soon see, that it may not be entirely true even at that level.

Of the many aspects of Hubble's theory of how far the Universe has expanded since the Big Bang, one of the most useful is the Hubble Parameter, written as $H0$. It is a measure of how rapidly space is expanding and of how old the Universe is. We calculate $H0$ using a large number of different yardsticks. We need to know how far a distant galaxy is, and its

Stephan's Quintet. A beautiful example of galaxies in which the redshift of one is markedly different from the redshifts common to the others. Is the anomolous one a member of the group or does it just appear that way? Or, perhaps there is a problem with our understanding of the redshift theory. (Photograph by Tim Hunter)

"velocity of recession," or how fast it appears to be receding, or moving away, from us. V. M. Slipher first observed that the spectra of these galaxies were shifted towards the red, and Hubble explained that this redshift could be used to indicate the amount of recession; the more extreme the redshift the greater the velocity of recession. Astronomers measure the velocity of recession from the redshift.

Where Redshifts Don't Work
Just as astronomers thought they had the redshift clue sewn up, strange objects began showing up that defied the redshift logic. Not far from the Great Square of the constellation Pegasus is a beautiful system of five galaxies called Stephan's Quintet. The five galaxies can be seen through large amateur telescopes. In long exposure images, they appear to be joined by a series of filaments, but by traditional interpretation of the redshift of one of the galaxies, it is only *one-eighth* as distant of the others.

Another example involves a quasar called Markarian 205. Although it appears to be very close to the galaxy NGC 4319 in the constellation of Draco, not far from the North Celestial Pole, its redshift indicates that it is *12 times* farther away! Is the galaxy's position simply a chance alignment between us and the much more remote quasar? Or, is the galaxy physically connected with the quasar, and the redshift due to some unknown process other than recession? We do not know the answer to this riddle.

The most startling example of "discordant redshifts" is in VV172, a chain of five galaxies of which one galaxy has a redshift that puts it more than twice as far away as the rest. If the lone galaxy's redshift is due to the expansion of the Universe, then this galaxy is much farther away, and by incredible coincidence it just happens to fill a gap in the chain. But if any of these rogue galaxies really do have redshifts that are caused by something other than Hubble's law, we need to find out what *does* cause it before we can have full confidence in what galactic redshifts tell us about how fast a galaxy is receding.

Determining the distance to a galaxy is another matter. Astronomers use a variety of yardsticks to calculate how far away a galaxy is: The Cepheid variable stars are by far the most accurate yardstick. Thanks to the Hubble Space Telescope, observers have been able to discover Cepheid variable stars in galaxies farther and farther away. These stars show a solid relationship between their average luminosities and the rate at which they vary: Measuring the cycle of variation of one of these stars is a fairly simple matter, since their cycles usually do not stretch beyond a few weeks. The brighter the average luminosity of the star, the longer its period is. A star's period thus gives a strong clue to how luminous it really is; comparing its period with its apparent mean brightness provides a measure of the star's distance.

Astronomers can determine the distances to more distant galaxies by using a method called the "Tully-Fisher relation." It works in the same way as the period-luminosity relation of the Cepheids, but relates the total luminosity of a galaxy to the rate of its rotation.

For galaxies much farther away, astronomers measure the rate of growth of the expanding shells of supernovae. A supernova shell's growth rate is then compared to its speed of expansion line-of-sight toward Earth. This method has been used on several supernovae in distant galaxies.

How Old is the Universe?
If we knew the value of the Hubble parameter, we would know the age of the Universe. The problem is that different yardsticks yield results that differ by as much as a factor or two. A recent long-term study of the sky, conducted by a group led by cosmologist Wendy Freedman, used the Hubble Space Telescope to examine the behavior of some

Cepheid variables in the spiral galaxies Messier 96 and Messier 100. Their conclusion: The Universe is as young as 8 billion to 10 billion years old. However, some white dwarf stars, common in our galaxy, can be easily dated at least six and as many as possibly fourteen billion years in age. And globular clusters appear to be even older. Since they orbit our galaxy on elongated elliptical paths, they should have been formed before the galaxy's disk shape evolved. The clusters can also be dated to at least 13 billion years old. So how can the Universe be younger than some of its stars? That cannot happen any more than a mother can be younger than her children. More recently, another orbiting satellite called the Wilkinson Microwave Anisotropy Probe (WMAP) also surveyed the Universe for fluctuations in the cosmic microwave background. One of the most important finds in cosmology came out of examining the data it collected. We can now say, with the greatest accuracy to date, that the Universe is approximately 13.7 billion years old. It will be interesting to see what errors in data or errors in its interpretation led to the earlier conclusion that some parts of the Universe are billions of years older.

Neither Beginning Nor End: A Steady State
The cosmology we have just explored is called the Big Bang, so named derisively by the British astronomer Fred Hoyle, who was part of a team that suggested a different way for the Universe to have formed and to function. They thought that the Universe exists in a "steady state," with matter being created out of nothing to make up for the Universe's continual expansion. It follows what they call the *perfect cosmological principle*, which states that not only is the Universe homogeneous in space, but it is also stable in time. The discovery of the quasars in 1963 offered strong evidence that the Universe has been evolving with time. Since all the quasars are very far away in space and time—all are in distant galaxies

whose light began streaming toward us long ago—it seems that there were more of them when the Universe was younger, and thus the content of the Universe has been changing. The discovery of the microwave background radiation, and especially its texture as observed by the COBE satellite, would indicate that despite some unanswered questions, the Big Bang cosmology seems to fit better the available evidence of observation.

The Great Attractor
We know that all the superclusters in the Universe are racing away from one another. But one interesting discovery shows that the galaxies in our neighborhood, including the entire Local Supercluster, seem to be heading toward a distant supercluster in Hydra and Centaurus. Composed of many thousands of galaxies, it is called the Great Attractor. The Attractor is some 200 million light years wide and is spread out over a large area in the southern sky. It contains the mass of more than 10,000 trillion suns, which is thirty times more massive than the Local Supercluster.

The Attractor is just one of many massive superclusters of galaxies. In 1936, Clyde Tombaugh detected one of the first known superclusters. Stretching from Pisces through Andromeda to Perseus, this vast array has recently been found to be associated with a second supercluster in the constellations of Ursa Major and Lynx. It is possible that the two groups are really joined, connected gravitationally in some way. If that is true, this "megacluster" stretches from horizon to horizon—across half the sky.

The Great Wall
In 1989, astronomers Margaret Geller and John Huchra of the Harvard-Smithsonian Center for Astrophysics reported yet another structure of galaxies, a huge conglomeration of at least 250 million and possibly as large as 500 million light

years across! Called the Great Wall, it stretches through the northern hemisphere spring sky. Geller and Huchra's method was to chart the distribution of galaxies in space in three dimensions, not the two that normal viewers of the sky are familiar with. Using this technique, Geller and Huchra have found that the superclusters lie on what looks like sheets atop bubbles. The largest "bubble" so far found is the Great Wall. In a different survey, the University of Hawaii's Brent Tully discovered that maybe 100 large and rich clusters of galaxies lie in a disk-shaped structure that is 1.5 billion light years long and 200 million light years wide.

In addition to the recent stunning finds of unimaginably large superclusters, astronomers are also finding large voids, where empty space stretches on virtually forever. Voids are not completely without matter, but the amount of matter within them is far less, perhaps a fifth of normal. If these structures are real, and not just statistical anomalies, then perhaps the Universe is not evenly distributed at the supercluster level, and we have not yet found the level at which even distribution begins.

The Problem of Missing Mass
Considering the variety of interesting objects we have looked at, from stars to clusters to galaxies to quasars, it seems incredible to have to report now that all these objects comprise only two-to-five percent of the estimated mass of the Universe! The problem comes from a simple calculation that there is not enough mass in what we can see in galaxies, including ours, that could hold these galaxies together in a cluster. Experiments repeatedly conducted on the Virgo cluster, of which we are nominally a part, show that the total amount of mass of stars and clouds that we can see is not enough to hold the cluster together. *There should be 50 times more mass.* The missing mass problem was first discovered when astronomers studied the motions of stars per-

pendicular to our galaxy's disk. If we take into account only the mass of stars and nebulae that we can see, these stars should be flying away from the galaxy's core; instead, they orbit it.

What form could this missing mass take? MACHOs, or Massive Astrophysical Compact Halo Objects, are one possibility. MACHOs are really brown dwarf stars or other stars too faint to be observed, and possibly these stars could be detected as they pass in front of brighter stars. As they pass in front of the stars, they would cause a gravitational lens effect, resulting in the bright star increasing in brightness for a time. Although many stars have been studied, only a few "microlensing" effects of this type have been observed.

Undoubtedly, the dark material should involve hydrogen and helium, and perhaps a few other light elements that were formed just after the Big Bang. But these light elements are also formed as a result of the fusion that goes on in stars, and when we study a cloud of gas we cannot distinguish between primordial elements and elements added during stellar nucleosynthesis.

Deuterium, or heavy hydrogen, is an isotope of hydrogen. Whereas hydrogen consists of a single proton around which an electron revolves, deuterium has a neutron, a particle without a charge, in its nucleus. Deuterium is not created in stars, so any deuterium we find must have been formed early in the Universe's life. The amount of deuterium, which is measured as the ratio of deuterium to ordinary hydrogen, is an important value because it reflects how dense the surrounding matter was at the time of its formation.

The Fate of the Universe
Deuterium can help tell us about the conditions in the Universe during its first few minutes. If it was very dense, then deuterium would have easily found available neutrons to join with and form helium. If the Universe was not so dense,

most of the deuterium would still survive. From the amount of deuterium found so far, it would seem that at its birth the Universe was not dense enough to cause its material to stop expanding some day, and then turn back, due to the force of its own gravity, and return to its point of origin.

The Universe will eventually end with dead stars in ghost galaxies expanding off forever. Eventually, stars in galaxies will plunge into central black holes, and atomic particles will decay. After more billions of years than we could count (one estimate is ten to the 100th power!), even quarks will vanish, leaving a little light, some leptons, and slowly evaporating black holes.

But what if there *is* sufficient mass, and therefore sufficient gravity, to slow the expansion to a full stop? Then the Universe will hold still for less than a second, and start a slow process of collapse. Redshifts will become blueshifts, and over billions of years the superclusters of galaxies will close in on each other. As they get very close to each other, temperature and pressure will rise until every atom of every star of every galaxy will dissolve into their nuclei. As the squeezing strengthens, nuclei will dissolve into protons and neutrons. Eventually these nucleic particles will dissolve into a soup of quarks and leptons. Other particles called "X bosons" will emerge near the end: a singularity of everything, and then nothing.

Olbers's Paradox

Some two centuries ago, physician Heinrich Olbers looked pensively at the night sky over Bremen, Germany and asked himself why it was dark. Olbers had observed that there are a certain number of bright stars, spaced at random intervals against the dark background, and a much higher number of fainter stars. With each increasing magnitude, in fact, there are far more stars. Should not the stars continue multiplying the fainter they get? Then why is the entire sky not blindingly bright with stars? If space is infinite and filled with stars, he suggested, then the entire sky should be as bright as the surface of the Sun. Obviously there are only a finite number of stars in our galaxy, but in the current age of our understanding of the Universe, the superclusters of galaxies should be filling the night sky with blinding radiation.

The answer to this old riddle, it seems, lies in the age of the Universe. If the Universe were several orders of magnitude older than it is—*hundreds* of billions of years old—and if we could see all the energy that ever existed within it, radiation from supercluster upon supercluster would indeed blind us out. But the Universe, no more than fourteen billion years old, is not old enough for light from its most distant objects to have reached us yet. Another factor lies in the expansion of the Universe. Because it expands, radiation from distant objects is redshifted and weakened, as every photon suffers a loss of energy during its shift to the red.

As each day ends, and we watch the Sun set, the world is left to darkness. At such a time we can ponder how incredible it is that this darkness is part of the way the Universe was born and continues to expand.

REFERENCES
Barrow, John, and Joseph Silk. *The Left Hand of Creation.* New York: Basic Books, 1983, 225-26.
Donne, John. "Goe, and Catch a Falling Starre." *Poems and Prose.* New York: Everyman's Library of Alfred Knopf, 1995. 14.

PART FOUR

THE MAGIC
OF ECLIPSES

CHAPTER 14

Childhood Impressions of a Darkened Sun

> A man's reach should exceed his grasp,
> Or what's a heaven for?
> —Robert Browning, *Andrea del Sarto*

A few hours ago, I completed observing session No. 11619 from our own Jarnac Observatory. The sky was warm and clear this May morning as I turned my telescope to the southeast to begin my pastime of searching for comets. As I moved the telescope from one field of view to the next, and on to the field after that, I searched a pattern of sky that might reveal a new comet. There were no new comets this time, but as I moved the telescope closer to the southeastern horizon, I thought of another observing session that took place almost exactly forty years earlier.

Now a distant memory, that observing session was No. 1 in my record book, my first formal, recorded observing session: *1S. October 2, 1959. Partial Solar Eclipse. Just last part observed because of clouds.*

CHILDHOOD IMPRESSIONS OF A DARKENED SUN

OCTOBER 2, 1959
Something magical happened early that morning as a partial eclipse of the Sun worked its way over my home in Montreal. Far to the south, the full darkness of the Moon's shadow cast its spell over a thin path through Massachusetts, and then out over the Atlantic. But I didn't care about that. For me, at age 11, all that mattered was that I was to see my first eclipse. The event was to last until 8 A.M., so that it would be conveniently over in time for the start of my day in sixth grade.

Our observing team consisted of my mother, brother Gerry, and me. As we drove to the Mount Royal Lookout facing Montreal's east side, we worried that clouds would prevent our seeing the eclipse. We waited for a while as the Sun rose behind the overcast sky. Then Mother noticed that the clouds were clearing from the west, so if we moved to another site farther west, we might get to see the end of the eclipse as the Sun broke through the clouds. We sped to the new site, got out of the car, and waited along with a large crowd that had gathered there. The sky grew brighter until the Sun peeked its way through the clouds. As the Sun rose further, its crescent shape showed itself at last. It was my first eclipse, my first observing session, and I was in heaven.

JULY 20, 1963
After that early eclipse, I looked up everything I could about eclipses. I used a book that contained a map of the world crisscrossed with thin lines that curved their way across the globe. I noticed that one of those lines showed the path of an eclipse that would cross Canada and rush near Montreal on July 20, 1963. It was a very curious thing, to see that the track of not just a partial but a *total eclipse* would be only a two-hour drive from downtown Montreal.

The summer of 1963, however, found me far from my Canadian home. My address was at Denver's Jewish National Home for Asthmatic Children, a spot from which I would see, at maximum eclipse, a half-covered Sun barely more exciting than my view from 1959.

Since it was the policy of the asthma home not to allow their patients to return home during their year-plus-long stay, my chances of seeing this eclipse seemed remote. But Mom and Dad put in a special request that I be permitted to return home for just one week to see the eclipse. The administration of the asthma home knew that under the clearer skies of the American West, my harmless little hobby had flowered into an all-consuming passion. I used every opportunity to observe the Sun by day, projecting the solar image on a piece of paper to count and draw the sunspots. Knowing how serious I was about observing the total eclipse, the asthma home permitted me to travel east.

The evening before the eclipse, we had dinner with our relatives and close friends Leo and Leona Kirschberg. Observing the total eclipse was on our minds, but Leo, an opthamologist, was adamant about taking care not to damage our eyes. He pointed out that during the partial phases of an eclipse, the Sun's visible light output drops, so that we are able to gaze at the Sun for longer periods without squinting. We also want to look at the changing Sun as the Moon covers more and more of its face. However, Leo warned, the ultraviolet rays coming from the Sun's photosphere are just as strong during an eclipse as they are at any other time, and these rays can permanently damage the retina.

I understood Leo's wisdom, but in reply I claimed that while this was true for the eclipse's partial phases, it was not true when the entire Sun was covered by the Moon, total eclipse, and by far the most interesting time to see the Sun. "But how," my father asked, "can one be sure that the entire Sun is covered?"

"The darkness is supposed to whoosh in like . . . like . . . "

"Like a rapidly approaching thunderstorm?" Leo asked, helpfully.

CHILDHOOD IMPRESSIONS OF A DARKENED SUN

"Much faster!" I said, having no understanding whatever of how vastly different the next day's event would be. In any case, we weren't sure we'd see this eclipse at all. "Chances for viewing tomorrow's eclipse in southwestern Quebec," the weather forecaster intoned, "are poor." Eclipse day dawned mostly cloudy, but we did see the Sun break through a few times while driving to our carefully selected site at Plessisville. Our group of four: Mom and Dad, my friend Paul Astrof, and I were on our way to be a part of a four-way lineup that included the Sun, Moon, Earth, and us.

The thing I remember most from the following day was Dad's astonished reaction to the fact that the eclipse started right on time. To astronomers used to traversing the globe to see an eclipse, that innocuous first contact of Moon and Sun is taken for granted. Dad was amazed that a centuries-old prediction of some long-gone astronomer was coming true before his eyes.

Though not a scientist, Dad was very good in seeing the poetry around him. As we prepared to watch the eclipse, we knew that at that moment thousands of other teams would be watching the eclipse as it tracked across the continent, from Alaska, all the way to us, in under two hours. So just at this moment another team would be watching the last thin crescent of the Sun disappear as the Moon pushed it into total eclipse. Braving mosquitos and a dismal weather forecast, the team consisted of planetary astronomers Brad Smith, Clyde Tombaugh, and his wife Patsy. The weather cleared in time for them to view a beautiful total eclipse. Clyde was already famous as the discoverer of Pluto, the ninth major planet of our solar system. Brad would become well known two decades later as the leader of *Voyager*'s imaging team, a project that led two intrepid spacecraft to explore, for the first time, the outer worlds of our solar system.

Clyde and Brad wanted everything observed, and timed, as accurately as possible. Accordingly, Clyde asked Patsy to keep her eyes on the stopwatch and count the 90 seconds of total eclipse. Obediently, Patsy did, and thus only got two quick glances of this eclipse. It was a memory she would carry until she finally would see her next total eclipse 35 years later.

The sky was clear in Alaska that eclipse morning, but late in the afternoon in southwestern Quebec it was mostly cloudy. The Sun broke through occasionally, but as it went deeper and deeper into partial eclipse, heavy clouds on the western horizon seemed sure to block our view of the total eclipse. In the last few darkening minutes before the onset of total eclipse, Dad looked up. "Come on, please give my son a break," he said quietly, "a break in the clouds."

Somehow his prayer was answered. The clouds held back to reveal the incredible sight of the onrushing shadow and a faint circular corona. With the sunspot cycle nearing its 11-year minimum, we did not expect to see any prominences. A minute later, the shadow lifted and a thin crescent of sunlight reestablished itself. Stunned and excited by what we saw, we just stood there. And then the clouds came, covering up the rest of the eclipse. But it didn't matter; we saw what we had come to see.

As we prepared to drive back to Montreal, the Moon's shadow continued its trek through southern Quebec and Maine, and out into the Atlantic. Near the end of its journey, it covered a swath of ocean some 350 kilometers southeast of Halifax. As we drove back that cloudy and happy afternoon, I had no idea that, 36 summers later, when the Moon's shadow would again be hitting the ocean at the same spot, I would be there, and I would see that eclipse again.

CHAPTER 15

All About Eclipses

Thy shadow, Earth, from Pole to Central Sea,
Now steals along upon the Moon's meek shine
In even monochrome and curving line
Of imperturbable serenity.
  —Thomas Hardy, 1903

Remember the old conundrum about the tree falling in the forest? If no one is there, does it make a sound? I like to apply that to eclipses, especially the effects that the 53 eclipses I've seen have had on me. I've always felt that we, as observers, are vital parts of these events. So suppose they gave an eclipse, and no one came?

As a scientific event, an eclipse is a comic coincidence, a curiosity. Planets don't crash into each other, and stars do not explode. But put yourself into the event, and an eclipse can have a most powerful effect. Even a barely noticeable penumbral lunar eclipse does that to me. The Moon's supposed to be full, but as its brightness dims and the rays stretching away from the craters Tycho and Copernicus become so much more prominent, I become aware that inexorably, the Moon is passing through the outer reaches of the shadow of the Earth. At the other extreme, of course, is a total eclipse of the Sun, an event that stabs like a knife to the core of my emotions. Sure, if no one sees an eclipse, then the event is nothing more than a coincidence. But those who do see it leave subtly changed, and moved by its power.

Let's look at the mechanism behind that power. A total eclipse of the Moon happens here on Earth, and with planets and moons elsewhere in the solar system. As the Moon orbits the Earth once every 29 days, it forms some angle between it, the Earth, and the Sun. Twice each month, at New and Full Moon, that angle becomes a straight line. If it is precisely straight, then an eclipse takes place. Eclipses of the Sun or the Moon can occur only when the Sun, Earth and the Moon are exactly lined up. This can occur during eclipse seasons, which happen twice each year. The simple geometry of planetary bodies orbiting one another in space is common enough, and because the solar system is essentially on one plane, like a record or disc, these lineups occur frequently.

But on other worlds the effect is far less dramatic. Mars, for example, has lineups when either of its tiny moons, Phobos and Deimos, pass in front of the Sun. But these moons are so small that they produce virtually no effect at all: If you were standing on the surface of Mars, you might barely detect such an event in progress as a tiny dot crossing the Sun's surface. Jupiter's moons would, on the other hand, block out the Sun so completely that they would cause several hours of darkness. But Earth's single Moon is now at just the right distance from us that it appears to be the same size as the Sun. The result is an exquisite blocking of the Sun's bright surface, revealing the prominences of the Sun's inner atmosphere, or chromosphere, and the pearly outer atmosphere, or corona. The fact that we get to see the prominences and the corona at all is a miracle in itself, for the Moon is slowly moving farther from Earth. As it continues to recede, its apparent size will shrink, and it will no longer be able completely to blot out the Sun.

Eclipses, Transits, and Occultations

Technically, the event called an eclipse takes place only when the body being covered is the same apparent size as the body doing the covering. A *transit* takes place when a small body passes in front of a larger one, as when Mercury or Venus pass in front of the Sun. An *occultation* is the term we use to describe the passing of a large body in front of an apparently smaller one. The Moon frequently passing in front of a star is an example.

As the Moon continues slowly to move away from Earth at a rate of a yard per century, it will—as noted before—eventually seem to take up less space in the sky. Some 620 million years in the future, there will come, sadly, a day when Earth will see the last total eclipse. The Moon will pass in front of the Sun, covering it entirely for a fraction of a second, and when that far-off event is over, total eclipses of the Sun will be a thing of the past. Even now, almost half the times that the Sun is centrally covered by the Moon, the Moon is near the farthest point of its orbit around the Earth. At that distance it does not cover the Sun; the result is an *annular*, or *ring* eclipse, at the middle of which the Moon is surrounded by a ring of bright sunlight.

In May 1984, I traveled to New Orleans to see an annular eclipse. A cold front had passed through the night before, leaving the usually humid city dry and clear. At the midpoint of the eclipse, the Moon's shadow swept out of the sky and almost entirely enveloped us. Overhead, what was left of the Sun shone as the thinnest of rings. A second later, the darkness whooshed away, leaving a thin crescent of sunlight. On the way home from this particular ring eclipse I stopped by to visit Clyde Tombaugh, the discoverer of Pluto, and an old and close friend. "How was the eclipse?" he asked. "Did you have a ringside seat?"

Clyde got the chance to live down that pun. Ten years later a second ring eclipse took place directly over his house. Although poor weather was forecast, the day dawned bright and clear and we saw a magnificent ring. The Moon was close to apogee, and took up considerably less space in the sky than the Sun. Thus, the ring was bigger, and it lasted 11 minutes!

The Orbit of the Moon
The moon circles the Earth once every 29.2 days, a period loosely coinciding with a month; the word "month," in fact, is derived from old German "moon." The Moon's orbit is the only determinant of the Moslem calendar, and is the prime base for the Jewish calendar, whose months, though timed to the Moon's orbit, are kept in step with the solar year by the occasional addition of a leap month.

Do the Moon's phases affect human behavior? There has been much conjecture and debate on that question. Anecdotal evidence does support the idea that people tend to get rowdier around the times of full Moon. But to the question on whether the Moon has an effect on the Earth, the answer is absolutely clear. When the Moon is near perigee, its gravity causes the tides to gain strength. And when Moon and Earth are approximately lined up with the Sun, tides increase.

The Nodes
Eclipses happen because of the relation between the orbit of the Earth around the Sun and the Moon's orbit around the

Earth. The Earth orbits the Sun in a near-circle once every year, and the Moon orbits the Earth, reaching the Sun's position, every 29.5 days. The two orbits are tilted relative to each other; if they were not, eclipses would happen every two weeks, a solar eclipse every new Moon and a lunar eclipse at every full Moon. Instead, the Moon spends part of its month-long orbit below the plane of the Earth's orbit around the Sun, and part of it above that plane. Twice each month the Moon crosses the plane of Earth's orbit. The two points of crossing, or intersection, of the two orbits, are called *nodes*.

The Moon crosses a node twice each month. If the Moon is moving northward in its orbit, it's called the *ascending node*; if it is going south, it crosses the *descending node*. The node crossings take place at different phases of the Moon each month. Now, in addition to the Moon's orbit of the Earth, the Earth travels round the Sun, so the Sun appears to cross one of the nodes twice each year. When that happens, an eclipse can occur.

On the Moon

When eclipses occur on Earth, does anything happen on the Moon? During an eclipse of the Moon, for example, the entire Moon is bathed in the shadow of the Earth, which means that a person or a camera on the Moon should witness an eclipse of the Sun. On April 24, 1967, humanity saw a total eclipse of the Sun by the Earth through the eyes of the U.S. space probe called *Surveyor 3*. The spacecraft's camera took two sets of pictures of the event. The exposures revealed that an eclipse of the Sun by the Earth, seen from the Moon, is less spectacular than one of the Sun by the Moon. At the same time that *Surveyor* was working to take those pictures of a solar eclipse, a lunar eclipse was taking place on Earth that Passover night. (Since they always occur at full Moon, Passover Seders often coincide with eclipses of the Moon. I remember leaving our Seder to observe this particular eclipse.)

REFERENCES
Hardy, Thomas. *The Collected Poems of Thomas Hardy*. New York: Macmillan, 1926.

CHAPTER 16

Of Cycles and Friends

> Contemplate all this work of Time,
> The giant laboring in his youth;
> Nor dream of human love and truth,
> As dying Nature's earth and lime . . .
> —Tennyson, *In Memoriam*

I saw the 1963 eclipse on July 20 for the first time. Although I didn't know it then, if those clouds came in too soon I would still have the chance to see the same eclipse again, on July 31, 1981, if I cared to travel to the Soviet Union. Eighteen years and a third of a month later, the Moon's shadow touched down a third of the world away, and crossed over the Soviet Union. I didn't see the eclipse then, but my colleague and friend Bart Bok was able to, thanks to an arrangement I had made for him. A famous specialist on our Milky Way Galaxy, Bart used the trip to see the eclipse and review his favorite topic—the bigger and better Milky Way—with his Soviet colleagues. A fierce promoter of international cooperation in science, he also used the trip to try to strengthen the bonds of cooperation during that dangerous Cold War period.

I missed the eclipse that time. But another eighteen summers later, the shadow would sweep out of the sky once more, touching the earth in the Atlantic Ocean some 350 kilometers southeast of Halifax, Nova Scotia. The date: August 11, 1999.

Saros 145: An Eclipse Through Time

The same eclipse, over and over again. Each eclipse repeats itself every 18 years, 10⅓ days (or 11⅓ days, when we take leap years into account.) The eclipses are almost identical, except that thanks to that extra ⅓ day, they fall over a different part of the Earth. We call each of these cycles of repetition a *saros*, which derives from a statement from an ancient astronomer named Suidas, that the length of "the saros" was 18½ years. Saros is the classical Babylon term for the number 3600, and hence refers to their period of 3600 years. However, to the ancient Chaldeans and Greeks, the saros had a different meaning—the period of 18 years, 10.3 days (or 6585⅓ days) during which eclipses would repeat themselves. The Chaldeans discovered that 6,585⅓ days after a new Moon at one of the nodes, the Moon has orbited the Earth precisely 223 times, and returned to the new Moon phase at the same node. At the same time, the Sun, in its apparent journey around the ecliptic, has passed through 19 eclipse years (of 346.62 days); it has also returned to the same node, and another eclipse occurs.

Summer of 1999: The Cycle Returns

In the predawn hours of August 11, 1999, saros 145 was about to return as I arrived at the forward promenade deck on the *Regal Empress*. The sky was clear and sparkling, a good sign for the total eclipse that would come with sunrise. Standing out near the bow, Mira, my orange telescope, in hand, I saw the ocean stretching out toward the horizon. Corona Borealis, high in the sky just a few hours ago, now sat low in the west. Toward the north, another vessel's lights shone in the distance. As it sailed northward, after half an hour its lights disappeared one by one as it dipped below the horizon to prove, once again, that Earth is round. There were no other ships, and no planes above, just the sea, the sky, and the moving ship.

Suddenly a faint flash of light brought me back to Earth. Startled, I wheeled around and looked up. Two decks above me, a bridge officer was lighting his cigarette. Enclosed by the windows surrounding the darkened bridge, the officer, responsible for directing the vessel's speed and bearing, might as well have been in a different universe; that's how dark and quiet it was on that deck, and that's how alone I was.

It was time to get busy. Standing there with the telescope, I felt like an old sea captain myself. Mira, however, was a lot more powerful than the small refractors that they had peered through to sight icebergs, land, and distant ships. But Mira had no interest in those earthly things. For the next hour, the objects I was searching for were not icebergs at sea but those in the sky.

Icebergs in the Sky

Suddenly realizing this relationship between comets and the sea, I felt that what I was about to do—search for comets aboard ship—was a totally justified task. As we saw in Chapter 2, comets actually consist of a combination of ices (not just water ice), and rock. On Earth, icebergs depart from the great ice fields in the far north and south, and then drift through the oceans—occasionally endangering passing ships as the passengers of *Titanic* found to their horror in 1912. Comets begin their journeys by leaving one of two large comet fields, the Kuiper Belt where Pluto lies, or the Oort cloud, much farther away. They drift through the ocean of space, where they can occasionally endanger passing planets, as the dinosaurs found to *their* horror 65 million years ago.

All this seems a cute analogy, but for me, on that morning, it was very real. As I stood on the bow of the ship, I stared out into two oceans, the Atlantic and the sky. Both stretched out as far as I could see, and as the ocean was so calm I could see stars and constellations reflected in it; in a sense, I couldn't even tell them apart. It was with this feeling of unity, calm, and utter aloneness that I pointed my telescope toward the northeast and began my search for icebergs in the sky.

A Comet Search

Searching for comets is a happy pastime that has occupied many of my clear evenings and mornings since I began the program on December 17, 1965. It's like being a night watchman. Because comets tend to be brightest when they are close to the Sun, I usually search the region of sky that is closest to where the Sun has set or where it will rise, and so, on this morning, I moved the telescope back and forth in horizontal sweeps across the northeastern sky. As we raced toward our rendezvous with the Moon's shadow in just two hours, I expected that the vessel's pitch and roll would make viewing through a telescope impossible. But that night's view was pristine: The sea was calm and the ship's motion resulted in only mild shifts of no more than a third of the field of view in the telescope's position as I swept slowly across the sky. Even better, northwesterly winds were following us, so there was virtually no wind up there at the vessel's bow. Since looking for comets is an activity that involves absolutely no mental effort, as my eye concentrated on what the telescope was showing me, field of view after field, I was free to contemplate the darkness of the night and the sublime meeting of two oceans.

Then I moved the telescope over one more field, and my brain slowly shifted its attention to the faint fuzzy object that was suddenly staring at me. This region of the sky, near Auriga the charioteer, is famous for its several clusters of stars, but the fuzzy object I was looking at was totally unfamiliar to me. For a moment the unlikely was a real possibility: just two hours before a total eclipse of the Sun, could I actually have discovered a new comet? As it turned out, I *was* looking at a comet, my first iceberg of this voyage. But it wasn't a new one. I later identified it as Comet Lee, a comet that had been in the evening sky and had recently moved far enough east to enter the morning sky. The comet incident added to the unique feeling of utter aloneness—not loneliness, but a feeling, indeed, of being alone on two infinite seas.

Sunrise

With the onset of dawn, my private reverie came to an end, and as the time for sunrise approached, the increase in sky brightness slowed down, then strangely stopped, and began to reverse—the sky began to get darker just before sunrise. Then a thin crescent Sun rose out of the sea, and as it cleared the horizon it just sat there, its bottom cusp resting on the waves like a sunbather relaxing and reading a newspaper while sitting on a beach chair. With the crescent now just a thin, curved line, everything was still—the sky, the sea, even the crowd. Some could actually start to see the shadow dropping out of the sky to the west. Through the special sun filters, the solar crescent was shrinking to just a spot. Just then Wendee removed her glasses to snap a picture of the Sun. She was about to replace them when she glanced toward the Sun.

"Ah David. David. DAVID!" she cried out. "Diamond Ring!" I ripped off my glasses and looked. The crescent was now just a bright point of sunlight, and the corona surrounding the Sun was bursting into view. The quiet on the ship ended in a burst of excitement as the Sun just vanished. In its place was a jeweled crown. My telescope was patiently waiting; if it was to get used, now was the time. I yanked its filter off and peered through, then silently gave it to Wendee.

"Oh David! There's prominences all over it!" Those ruby prominences were the highlight of the view through the scope, she said later. We could see them for the tongues of flame they were, each one arcing majestically away from the Sun, each one fully capable of swallowing the Earth. I've seen these before during eclipses but never so large or so ruby-red; their closeness to the horizon must have intensified their color.

While the rest of the group was looking through the telescope, I glanced toward the horizon again. The seconds were ticking by. The total eclipse utterly dominated the view, but it didn't complete the view. Surrounding the Sun, and extending more than halfway around the sky, was a cigar-shaped ribbon of darkness that was the shadow of the Moon, more clearly visible in the dark sky than I had ever seen it before. But it was moving fast, rocketing past us at more than twelve thousand miles per hour. Like the second hand of a great clock, the shadow spun around to the northeast. Very little of it remained southwest of the Sun, and it was moving away fast. I also noticed that, looking out to sea, all was incredibly quiet. There were no planes overhead, and no ships around us. No one. Just us, the sea, the sky, and the incredible spectacle of a total eclipse of the Sun.

REFERENCES
Houghton, Walter E., and G. Robert Stange. *Victorian Poetry and Poetics*. Boston: Houghton Mifflin, 1979, 79.

CHAPTER 17

Sun, Moon, and Surprise

" 'The moving finger writes and, having writ, moves on.'
Like a moving finger of darkness the cone-shaped shadow of the moon had dipped down, scrawled its brief two minute
mark of night across the land and then moved on, still writing,
but now with invisible ink upon the empty page of space."
—Leslie C. Peltier, *Starlight Nights*

When the Moon passes in front of the Sun, cutting off its light, the result is one of the most spectacular sights Nature has to offer. Day turns into twilight as the shadow of the Moon cuts across land and sea, and the Sun appears as a jeweled crown hanging in the sky. But if all this isn't enough, the alignment of Sun, Moon, and Earth has other effects as well.

Twice each month, whether or not an eclipse takes place, the Sun, Earth, and Moon are aligned in such a way that tides on Earth are stronger than usual. Although this effect is most noticeable with the ocean tides, it is felt by the Earth's rocky crust as well. It might be just a coincidence that the great Tokyo earthquake of September 1, 1923, in which 140,000 people were killed, took place just 5 days after the August 26

COSMOLOGY 101

Twice each day, at the eastern end of Canada's Bay of Fundy, a flow of water equal to the combined currents of all the rivers on Earth passes through the Minas Channel into the Minas Basin. In these two photos (above and opposite), the combination of high astronomical tides and a low pressure atmospheric system produced extreme low and high tides.

SUN, MOON, AND SURPRISE

The photos are taken at Hantsport Pier, Nova Scotia. Note how, in the second picture above, the dock has completely disappeared.

Gravitational lensing. The Abell 2218 galaxy cluster is gravitationally lensing the light from much more distant galaxies. The "arcs" and "smudges" are images of distant galaxies that have been "bent" by the gravity of the nearer ones. (Courtesy NASA/HST and W. Couch, UNSW.)

SUN, MOON, AND SURPRISE

partial eclipse of the Moon, and the Turkey earthquake of 1999 occurred just five days after a total solar eclipse passed over the same location.

Astronomical Tides in the Minas Basin
The higher than normal tides that accompany these alignments of Sun and Moon are called astronomical tides, and if a low pressure area is centered over the area at the same time, they can be truly remarkable. As mentioned in Chapter 1, on the shores of Nova Scotia's Minas Basin these tides are amplified by the unusual resonance effect of the Bay of Fundy. Resonance works like a child on a swing, who can be hurled higher and higher if someone pushes the swing each time it reaches a limit. But these effects can be enhanced when the Sun and Moon are pulling together. On January 29, 1979, physicist Roy Bishop of Acadia University compared low and high tides at Hantsport, Nova Scotia, on a day near new Moon (in fact a month before the February 26 solar eclipse), and on a day when a low pressure system was passing through the area. His goal was to obtain photographs from the same location, the edge of the Hantsport pier, at low and high tide. The low tide photograph was easy enough to take; he drove his car out to the edge of the pier (he wanted the car in the picture for scale), then he walked back to take the picture. The wooden mounts for the pier were fully visible in the mud that cloudy morning. When he returned six hours later, he was astonished to find . . . nothing. The mounts for the pier were gone, buried under water, as well as the pier itself! The water level had risen by more than 50 feet. Roy couldn't drive his car onto the pier; in fact, to work his way to the same spot he had to creep, step by careful step, across the submerged planks, snap the second picture, then carefully work his way back. The result is a graphic demonstration of the influence of the Moon's gravity.

SUN, MOON, AND SURPRISE

Surprising Science Discoveries

When the Moon eclipses the Sun, Earthbound astronomers have a golden opportunity to study the star that gives us life. Since eclipses provide the only opportunities to see the Sun's corona from the surface of the Earth, scientists study the corona at these rare opportunities. They try to understand why it is so hot; far hotter than the surface of the Sun; the corona shimmers at some 2 million degrees Celsius. We study it to learn how it affects the Earth as well; its outer reaches, in fact, are the "solar wind" of radiation that reaches Earth.

The Eclipse of August 18, 1868, excited several groups of scientists. On that day a group of scientists, including John Herschel, great-grandson of William, discovered that the solar prominences were composed of hydrogen gas. At that same eclipse J. Norman Lockyer and Pierre Janssen detected a yellow line in the spectrum of the Sun's corona that signified a new element, which they named Helium, a quarter-century before it would be detected on Earth.

Comets have even been found during total solar eclipses. Arthur Schuster photographed a new comet during the total eclipse of May 17, 1882. (Littmann, Willcox, and Espenak 157). People enjoying the eclipse of November 1, 1948, to cite a recent example, were stunned to see a comet brighter than Jupiter about a Moon in diameter from the eclipsed Sun, and with a tail that stretched toward the horizon. (Kronk 147)

The eclipse of June 8, 1918, became famous for something else that happened on the same day. As Comet hunter Leslie Peltier wrote in *Starlight Nights,* his biography:

> When darkness came that evening I clamped my spyglass to the grindstone mount which still was standing at the station underneath the walnut tree. I hoisted it up on my shoulder and carried it out into

the middle of the front yard and stood it where I would have a clear view of the variable stars in the southeastern sky. That was the night that I forgot all about telescopes and variables for as I turned and looked up at the sky, right there in front me squarely in the center of the Milky Way, was a bright and blazing star!

The discovery of an exploding sun, or nova, is still one of the most exciting things that can happen in astronomy, and it was an incredible thing that such a star's light would have traveled for thousands of years through space to arrive on Earth's doorstep just as the Sun was being eclipsed.

REFERENCES

Kronk, Gary. *Comets: A Descriptive Catalog*. Hillside, N.J.: Enslow, 1984.

Littmann, Mark, Ken Willcox, and Fred Espenak. *Totality: Eclipses of the Sun*. Oxford: Oxford U P, 1999.

Peltier, Leslie C. *Starlight Nights: The Adventures of a Star Gazer*. 3rd ed. Cambridge: Sky Publishing, 1999.

CHAPTER 18

The Power of Gravity

If the Moon's gravity, aided somewhat by the gravitational pull of the Sun, is strong enough to affect the tides on Earth, then the gravity of the much larger Sun is strong enough to control the orbits of the planets, including Earth and Moon, and the eclipses they produce. And if Einstein's theory of relativity is right, then a beam of light from a distant star will be bent by the curvature of space as it passes near the Sun.

What does this bending of light have to do with eclipses? The answer is a story that began in the 1890s, when a teenaged Albert Einstein had a thought. "What would the world look like," he mused, "if I rode on a beam of light?" The answer, he later figured, was that the planet would be frozen in time, its clocks still, its action caught as in a photograph. A decade after he first thought about that, the young physicist, then unable to find a job in physics, began work at the Swiss patent office. Einstein's responsibilities there were not particularly time consuming, and he had time to ponder questions about physics, like the relation between matter and energy.

The result was Einstein's special theory of relativity, which appeared in a 1905 article without any reference or citation—virtually unheard of for a research paper. This was a completely original piece of work. Later that year, Einstein attached an additional thought to that article, the tiny equation $E=mc^2$. In those simple letters lay the idea that mass and energy are equivalent. Late in 1915, Einstein's general theory of relativity offered a new definition of gravitation that related it to space and time. In Einstein's physics, gravity is not a force but geometry. As any object moves, whether it is a baseball, a planet, or a star, it follows a geometric path shaped by the unified effect of mass and energy. Newton invoked a force of gravity to make his Universe work, and in almost all cases, Newton's laws fit. But where there is a lot of matter, like a star, Newton's laws fail, and it is important to see gravity not as a force of Newton but as Einstein's geometry of space and time.

How can this beautiful, simple relation be tested? In the years between 1905 and 1919, Einstein's theory caught the attention of physicists, but many thought that it was unprovable. Such a theory would be forever relegated to the backwaters of physics as a curiosity. In 1918, the British astronomer Arthur Eddington wrote the first English language account of relativity, and he noted how the theory was successful in solving the old problem about Mercury's orbit. As Mercury circles the Sun at a distance of only 36 million miles, its perihelion, or closest point in its orbit to the Sun, shifts a small amount with each orbit. The entire orbit is precessing by 43 seconds of arc per century, a tiny amount to be sure, but one which Newton's theory of gravitation could not explain. If Newton's laws were right, another planet closer to the Sun must be affecting Mercury's orbit. Scientists searched for such a world for decades, and even named it Vulcan. There is no such planet. Instead, Newton's model of gravitation turned out to be insufficient. But Einstein's theory explained this shift.

The Eclipse of May 29, 1919

Eddington noted that Einstein's theory "further leads to interesting conclusions with regard to the deflection of light by a gravitational field," and that it could be tested through an experiment. In 1918, Eddington was certain just what experiment would work. By photographing a star near the Sun, and then comparing its position with that on other photographs taken when the star is far from the Sun, Einstein's relation could be tested. Although this proposed test is fine in theory, when the star's light passes that close to the Sun, we shouldn't be able to sight the star at all! Except, that is, during an eclipse.

Just what was this result? Eddington needed to answer two questions: First, does light have weight, as Newton suggests? And if the answer is yes, is the amount of deflection in agreement with Newton, or with Einstein? The deflection, incidently, is in the opposite direction from the body doing the deflecting. The star will appear to be displaced outward, or away from the Sun, by the same amount as the total deflection.

For the experiment to work, Eddington needed to use distant stars, not planets, asteroids, or comets within the solar system. The bending of the star's light is detectable only with stars apparently near the Sun, and these stars can be seen only during a total eclipse. But the Sun's corona is also bright, so the only eclipse that will work is one when the Sun is near a group of moderately bright stars. It turned out that just such a coincidence was about to happen, and Eddington developed a plan to photograph stars in the Hyades star cluster, from Principe, a small island off Africa's west coast, which on May 29, 1919, would be under the darkness of a total eclipse of the Sun. The cluster would be just south of the eclipsed Sun, its stars bright enough to be captured on the photographic films of the time.

The displacement of a star close to the Sun is measured in comparison with stars that are farther from the Sun and not displaced. In order to measure displacement, astrometrists (astronomers who measure the positions of stars) therefore need to observe stars close to and farther from the Sun. Eddington wrote:

> In a superstitious age, a natural philosopher wishing to perform an important experiment would consult an astrologer to ascertain an auspicious moment for the trial. With better reason, an astronomer today consulting the stars would announce that the most favorable day of the year for weighing light is May 29. The reason is that the sun in its annual journey round the ecliptic goes through fields of stars of varying richness, but on May 29 it is in the midst of a quite exceptional patch of bright stars—part of the Hyades—by far the best star-field encountered. Now if this problem had been put forward at some other period of history, it might have been necessary to wait some thousands of years for a total eclipse of the sun to happen on the lucky date. But by strange good fortune an eclipse did happen on May 29, 1919. Owing to the curious sequence of eclipses [the Metonic cycle we have already discussed] a similar opportunity will recur in 1938; we are in the midst of the most favorable cycle. It is not suggested that it is impossible to make the test at other eclipses; but the work will necessarily be more difficult.

Planning an expedition to Africa to observe an eclipse in order to confirm the ideas of a German scientist was an almost insurmountable problem, but Sir Frank Dyson, then England's Astronomer Royal, was able to persuade the government that this eclipse presented a rare opportunity indeed, and that they should spend 1,000 pounds for an expedition to test the theory of relativity.

An Amazing Expedition
Five months before the expedition began, Eddington photographed the Hyades field, using the same telescope as would be brought to Africa. With the Hyades far from the Sun, this photograph would serve as a base for comparison. As summarized in A. Vibert Douglas's biography of Arthur Stanley Eddington, the night before the expedition set sail, there was a discussion about just how much deflection the star would suffer. If the deflection was a tiny 0.87 arcseconds, then it would confirm Newton's classical theory of gravitation. If it were much greater than that, or 1.75 arcseconds, then Einstein's theory would be confirmed. On the evening before the sailing began, Cottingham, who was to accompany Eddington, asked jokingly what would happen if the star's deflection was double what Einstein had predicted. Dyson, the man who had planned and arranged funding for the expedition, replied, "Then Eddington will go mad and you will have to come home alone!"

Eddington described this fateful expedition in his notebook: "We sailed early in March to Lisbon. At Frunchal we saw [Davidson and Crommelin, the other expedition] off to Brazil on March 16, but we had to remain until April 9 . . . and got our first sight of Principe in the morning of April 23 . . . about May 16 we had no difficulty in getting to check photographs on three different nights. I had a great deal of work measuring these."

The group arrived in Principe with a 13-inch diameter, 11-foot, 4-inch long refractor. They stopped the lens down to 8 inches to improve the sharpness of its images. The telescope was mounted in a fixed position, and a mirror, or coelostat, directed the light from stars into the telescope.

Douglas wrote:

> "On May 29 a tremendous rainstorm came on. The rain stopped about noon and about 1:30 when the partial phase was well advanced, we began to get a glimpse of the sun. We had to carry out our programme of photographs in faith. I did not see the eclipse, being too busy changing plates, except for one glance to make sure it had begun and another half-way through to see how much cloud there was. We took 16 photographs. They are all good of the sun, showing a very remarkable prominence; but the cloud has interfered with the star images. The last six photographs show a few images which I hope will give us what we need...."

A year later Eddington expanded on the details of those few minutes of total eclipse: "There was nothing for it but to carry out the arranged program and hope for the best. One observer was kept occupied changing the plates in rapid succession, whilst the other [presumably Eddington himself] gave the exposures of the required length with a screen held in front of the object glass to avoid shaking the telescope in any way."

Eddington's 16 exposures ranged in time from 2 to 20 seconds. The first ones did not record any stars, but they did capture the prominence. Happily the clouds cleared more toward the end of totality, and one photographic plate recorded five stars. Once all the pictures had been processed, Eddington made his first measurements at the eclipse site a few days after the event. The two photos were placed "film to film" in the measuring machine so that the star images were

THE POWER OF GRAVITY

close to identical. "In comparing two plates," Eddington wrote, various allowances had to be made for refraction, abberation [of the telescope lens], plate-orientation, etc. (Eddington 114-15).

Again from Eddington's log: "June 3. We developed the photographs, 2 each night for 6 nights after the eclipse, and I spent the whole day measuring. The cloudy weather upset my plans and I had to treat the measures in a different way from what I intended, consequently I have not been able to make any preliminary announcement of the result. But the one plate that I measured gave a result agreeing with Einstein." (Douglas 39).

As he completed the reduction of this plate, Eddington realized the significance of his result. Turning to his colleague, he smiled and said, "Cottingham, you won't have to go home alone." They packed their precious plates and returned to England. Four additional plates, of a different type, which could not be developed in the hot African climate, were developed there, and one of them confirmed the result shown on the first successful plate.

For a result as crucial as this one, Eddington had to make sure that instrument errors could not have led to the result. As a check, Eddington photograped a different star field, at night, with his arrangement at Principe; the field was photographed from England. If the "Einstein deflection" were the result of a telescope error of some kind, it would have turned up in these check plates. But no changes were found in the stars on these check plates.

The Brazil part of the expedition had much better luck. Their weather on eclipse day was superb, and they remained at their site for two additional months in order to photograph the same region of sky under cover of morning darkness. They had two telescopes, one similar to the African scope, and a much longer, 19 foot long, 4-inch diameter refractor.

But despite the best preparations, few observing expeditions are free of surprise, uncertainty, and disappointment. When the Brazil expedition returned home finally, its measured deflections with the larger telescope did *not* agree with Einstein, but with Newton! This was a shocking result, one which would inevitably delay any announcement, one way or another, about Relativity. Eddington suspected that the Sun's rays in the clear Brazil sky might have distorted the coelostat; in this one case, bad weather helped: "at Principe," he wrote, "there could be no evil effects from the sun's rays on the mirror, for the sun had withdrawn all too shyly behind the veil of cloud." (Eddington 117).

The final verdict on Relativity thus had to wait until a measuring engine could be modified to accept the seven oddly-sized plates taken through Sobral's 4-inch refractor. These plates seemed ideal, and their images were perfect. So was the result they revealed: a deflection in strong agreement with the results in Africa, and one in favor of Einstein's theory.

Einstein was thrilled with this result. "I should like to congratulate you on the success of this difficult expedition," he wrote in a letter that began "Lieber Herr Eddington! I am amazed at the interest which my English colleagues have taken in the theory in spite of its difficulty. . . . If it were proved that this effect does not exist in nature, then the whole theory would have to be abandoned" (Eddington 40-41). Einstein was right, and his theory of Relativity made the front pages of newspapers around the world. The man in the patent office had completely redrawn our understanding of the structure of the Universe, and his idea was proved correct thanks to a total eclipse of the Sun. Perhaps it is the special magic of solar eclipses that makes us want to see something superb, something that utterly blows us away, come out of them. In their discovery that the Sun's mass bent the light from nearby stars, Eddington and his colleagues helped Einstein give us a new universe, or at least a completely new understanding about the old one. That is quite a thought to ponder when you next have the opportunity to watch an eclipse of the Sun.

REFERENCES

Douglas, A. Vibert. *The Life of Arthur Stanley Eddington*. London: Thomas Nelson and sons, 1956.

Eddington, Sir Arthur. *Space, Time, and Gravitation: An Outline of General Relativity Theory*. Cambridge: Cambridge University P., 1920.

CHAPTER 19

The Eclipse Experience

> O dark, dark, dark, amid the blaze of noon,
> Irrecoverably dark, total eclipse
> Without all hope of day!
> —John Milton, *Samson Agonistes*

No matter how many eclipses I might see, I am always amazed at the uniqueness of each experience. As I was soon to learn, Bart Bok was right about seeing his 1927 eclipse—even a clouded-out eclipse is something to remember. Eclipses seen on snow, over the ocean, alone, or with others, all work their special magic on those who are fortunate enough to observe them.

1970

In the spring of 1970 I was a student at Acadia University in Wolfville, Nova Scotia. Knowing that the Moon's shadow would trace a path up the Atlantic seaboard that March 7, and pass over Nova Scotia about an hour's drive south of Acadia, I joined a group of friends to catch the eclipse's total

THE ECLIPSE EXPERIENCE

phase. We were clouded out, but we did see the Moon's dark shadow race across the clouds.

On that same day, Roy Bishop was smarter than I was. A physics professor at Acadia, Roy was starting to expand his interest in astronomy. Within a decade he would be National President of the Royal Astronomical Society of Canada, and editor of its *Observer's Handbook*. Back in 1970, Roy also wanted to view the eclipse. On seeing the same clouds I did, he checked a weather map in the local newspaper that day and thought that clearing should be coming from the west, but a little slower than had been forecast. He turned the pages of his phone book until he found listings for a town about 100 miles to the west. He dialed the number of a flower shop, and when they answered, he said, "Hello, this is Roy Bishop of the Physics Department at Acadia. Is the Sun shining there?"

"Why, yes, it is!" came the answer. Roy's group drove out there. "Words cannot describe the beauty of the corona and darkened sky I saw," he wrote to the Canadian Broadcasting Corporation the next day. Although I missed seeing that eclipse directly, I will never forget how the landscape was thrust into a darkness more profound that at any other eclipse I have seen since.

1979

It had been many, many years since the 1963 eclipse, and I really wanted to see another one under a clear sky. That opportunity came on February 26, 1979, the last total eclipse visible over the continental United States until 2017. I was in Canada, though, set up near Lundar, Manitoba. The sky was supposed to be cloud-filled that day, but the night before the eclipse a weak high pressure system formed over southern Manitoba, clearing the atmosphere and giving us a beautiful view of the corona and several spectacular prominences.

Solar eclipse: moon shadow. After it left us, the shadow raced across the Atlantic and over Europe. This photo is from the European

1991

In 1991, I saw an eclipse that was part of mighty saros cycle 136, the producer of one of the longest and finest series of eclipses ever seen. This saros was responsible for the famous

THE ECLIPSE EXPERIENCE

Meteosat 6, taken at 22,000 miles up. (Image copyright 2001 EUMES-TAT.)

Einstein eclipse of 1919, and was returning for an even better one. This one was going to be a rare eclipse that crossed the site of the world's largest observatory, whose telescopes were poised to study the eclipse as never before.

In Hawaii's early morning, the total eclipse would last almost four minutes. But where I was, with a *Sky & Telescope* group in Mexico's province of Baja California, the eclipse would culminate near high noon where totality would be longest, very close to *seven minutes* of darkness!

For the first half hour it was hard to imagine that anything unusual was taking place. The sky was still bright, and the beach was as crowded as any beach would be on a holiday afternoon. The sky did not darken linearly and steadily, as it does after sunset. My first thought that the light was changing came as I noticed that the sky was not quite as bright as should be at high noon in summer. Through my welder's glass I looked and saw that fully half the Sun was gone. By the time the sun was three-quarters covered the pace of darkening was increasing almost asymptotically. I looked away from the sea toward our hotel, which now towered into a navy blue sky. We looked at a hundred crescents projected by spaces between leaves, by breaks in a straw hat, between the crossed fingers of two hands—the crescents were everywhere.

The sky was now darkening fast. Cameras were being reloaded in these final minutes. A man from the bar approached, armed with bottles. "Corona beer?" he inquired hopefully. Now there was just a sliver of the Sun left, and the Moon was coming so fast I could see the Sun shrink as the seconds ticked away. Directly to the west the dark shadow of the Moon gained substance. Now the darkness was coming in waves! Was this my imagination, or was I seeing some sort of shadow band effect all around me, as the sky got two steps darker, then one step brighter, then three steps darker again. I had seen this effect in Manitoba at the 1979 eclipse.

Then I turned around and looked toward the vanished Sun. Stretching some three solar diameters east and west, through the telescope it was rich with streamers and intricate brushes of light. On the north and south sides were a

THE ECLIPSE EXPERIENCE

series of smaller eruptions of rays, shining outward like the mouths of baby birds in a nest.

Totality lasted long enough to allow me to use my telescope in a brief search for comets. My telescope passed over the third magnitude star Delta Geminorum, almost lost in the Sun's corona. I realized that we were witness to a strange coincidence of history: The eclipse was taking place directly over the spot where Pluto was when Clyde Tombaugh found it over sixty years ago!

Eclipses tend to heighten the senses. Listening to the sounds of the eclipse—the birds flying and strutting about, the incessant clicks of a million cameras—all accompanied the eerie light that surrounded us. Despite the fact that we were deep in the Moon's shadow, the sky was bright as twilight and landscape features were plain. Around the horizon was a bright red glow.

Other observers reported strange environmental effects. At her site, astronomer Jean Mueller reported several cows "coming home," single file into their enclosure. And down the beach from us, someone—of course not from our group—decided to show it all at mid-eclipse. With the intricate structure of the corona just hanging at the zenith, time seemed to stop. However, there were complex changes occurring all around. As we grew deeper into shadow the horizons continued to darken slightly. As the Moon moved off the west side of the inner corona a glorious orange prominence appeared. Through the telescope the prominence was so brilliant it was difficult to look at, and it seemed to have erupted off the disk of the Sun, just hanging there in space.

In the final seconds of totality the western sky started to brighten rapidly. I looked back at the Sun, admiring the prominence as the western limb brightened quickly. Suddenly a sharp speck of solar photosphere stabbed through the darkness, slowly spread out into a thin crescent, and then the eclipse was over.

1998

Seven years later, I saw my next total eclipse, this one aboard the *Dawn Princess* in the Caribbean. On February 26, 1998, this eclipse was the next one in the Metonic cycle following the one I saw in the frozen Canadian prairie in 1979. This time we were in a much warmer climate.

Although I had now been under the Moon's shadow eight times, five for total eclipses and three for annular eclipses, I was still not prepared for the splendor beyond words that the experience offers. And this eclipse, far from being just another eclipse, was the most precious of them all, for it was Wendee's first total. The onset of darkness came very gradually at first; with half the Sun gone, we realized that sunglasses were no longer necessary. But as the minutes rushed by, the gathering darkness began to hit us head-on. And as the crescent Sun shrank to a sliver, I tried to let Wendee experience as much of the event as she could.

"Glass on," I said, "now look at the Sun!" With her welders' glass, Wendee saw that the sliver was getting smaller. "Now look away from the Sun, and remove the glass." Venus had just popped out of the weird twilight that was engulfing the ship. Looking westward, I asked her to notice the distant ribbon of darkness that was the onrushing shadow of the Moon. "Glass on, and look at the Sun!" The sliver was little more than a line of light. "Glass off, look at the shadow!" In a matter of seconds, that distant ribbon was gaining strength and power. "Glass on!" The Sun was a short line. "Glass off!" The shadow was rushing at us fast. "Glass on!" The Sun was a point of light. "Now Wendee, keep looking at the Sun, and glass off." The diamond ring was beyond expression, a bright spark of sunlight and a thin corona. Within two more seconds, the diamond itself gave way to the Moon, and the corona swelled in size. Mercury, Venus, Mars, and Jupiter, all these planets hung in the sky, servants to a Sun whose light had been quenched. For just a few minutes, time stopped as Nature put on its truly magnificent show.

Each eclipse experience is different. That thought should be obvious, but I never fully realized it until after I saw the 1998 eclipse with Wendee. After 1979 and 1991, I honestly didn't feel I needed to see another total eclipse. After 1998, however, I was totally hooked—I didn't want to miss another one. In chapter 16, I described the 1999 eclipse. In 2001, I saw yet another in Zambia. The unique thing about that eclipse was the sheer length of time I could see the corona. By blocking the Sun behind a tree in the minutes after totality, I glimpsed the corona for 6 more minutes, bringing the total corona view for that eclipse to some 9 minutes!

REFERENCES

John Milton, "Samson Agonistes," (line 80) *The Norton Anthology of English Literature,* ed. Meyer Howard Abrams (New York: W.W. Norton, 1968) 706.

PART FIVE

TELESCOPES AND OBSERVING

CHAPTER 20

Telescopes

I succeeded in obtaining permission to view your sentence, the reading of which, though on the one hand it grieved me wretchedly, on the other hand it thrilled me to have seen it and found in it a means of being able to do you good, Sire, in some very small way; that is by taking upon myself the obligation you have to recite each week the seven psalms, and I have already begun to fulfill this requirement and to do so with great zest, first because I believe that prayer accompanied by the claim of obedience to Holy Church is effective, and then, too, to relieve you of this care. Therefore had I been able to substitute myself in the rest of your punishment, most willingly would I elect a prison even straiter than the one in which I dwell, if by so doing I could set you at liberty.

                        Suor Maria Celeste to her father, Galileo,
                                          October 3, 1633

The love and respect of a daughter to her father is so beautifully expressed in Sister Maria Celeste's haunting letter to her father on the occasion of her reading the sentence pronounced against him. In addition to a prison sentence that was later changed to a form of house arrest, Galileo was forced to recite Psalms 6, 32, 38, 51, 102, 130,

TELESCOPES

and 143—the seven penitential psalms once each week for a duration of three years (Sobel 2001). Sister Maria Celeste's commitment to recite these psalms on her father's behalf ties her to her father's plight, and to the great importance of his work to which she alludes in her Latin thought.

Galileo's works were already well advanced by the beginning of the evening of January 10, 1610, possibly the most important night in the history of astronomy, when he turned his telescope to the heavens and discovered the moons of Jupiter. From Galileo's first spyglass to the mighty Hubble Space Telescope to the adaptive optics that are used in today's most advanced groundbased telescopes, optics helped us reach the worlds, stars, galaxies, and other objects we have explored in this book. The telescope's epic journey might have begun as early as the year 1270, when Roger Bacon allegedly described a telescopic device. But did Roger Bacon invent a telescope? Although he was a public figure whose activities were known, there is no evidence that he actually built one—but that was more than 700 years ago. The first recorded construction of a telescope clearly took place in Holland, where Hans Lippershey, a maker of spectacles from Middelberg, in Zeeland, built one during the summer of 1608. He accomplished this by pointing two spectacle lenses in line with each other, and seeing the result as a magnification of distant objects. It is possible that he got the idea from his children as they played with two of his lenses. On October 2, 1608, Lippershey petitioned for a 30-year patent to construct a device for seeing at a distance (Bell 1981). Just two days later the States General, governing body of the Netherlands, who considered such patent requests, appointed a committee to test the instrument from the Tower at the Palace of St. Maurice. The test was very successful and Lippershey was asked to construct a binocular version of his device. Unfortunately, because the invention was known at the time to others, the States General denied the patent request. Lippershey, it seemed, did not keep his own invention a secret during that summer, so that by the time of his petition, others were making telescopes as well. Thus, although we do not know who first put two lenses together to make the very first telescope, we do suspect that Lippershey was the first to make a public record of it.

Galileo's Telescope

By early 1609, Lippershey's invention had spread far and wide, and foot-long telescopes were for sale in Paris shops. Word spread as far south as Italy, where it attracted the attention of a genius of the highest order, named Galileo Galilei. A remarkable man, Galileo Galilei was born in 1564, the same year as Shakespeare, in the Italian city of Pisa. The duplication of his first and last names, incidentally, is a Tuscan custom common to first born children (de Santillana 1955). Although he lived in a time when most thinkers believed in an Earth-centered Universe, he read Copernicus's Sun-centered theory in *De Revolutionibus* with enthusiasm. When Galileo learned of the invention of the spyglass, he promptly set out, in the summer of 1609, to design and build one for himself. This telescope was a *refractor*, which consisted of a lens held near the eye (we now call that the *eyepiece*) and another lens at the opposite end of the tube. That lens, called the *objective*, is responsible for gathering the light and sending it to the eyepiece. The Venetian Senators, very impressed with this device, pointed it toward the Mediterranean. It could spot ships coming before the unaided eye could.

Increasing the telescope's power to ten, Galileo next turned his attention to the heavens, where one of his earliest observations was, naturally, the Moon. He noticed that as the Sun rose over portions of the Moon, the peaks lit up with sunlight before the valleys, just as happens on Earth. His most historic observation was the discovery of Jupiter's four moons. These worlds did not revolve about the Earth, and thus they strongly suggested that Earth was not the center of the Universe. Although Galileo may not have been the first to observe these moons—Simon Marius, of Belgium, might have observed three of them a year earlier—Galileo was clearly the first to understand their significance. In March of 1610, Galileo wrote in a publication called *Nuncius Sidereus* (*Starry Messenger*) of both the existence of his telescope and the discoveries which it had wrought: "On the seventh day of January in this present year 1610, at the first hour of night, when I was viewing the heavenly bodies with a telescope, Jupiter presented itself to me; and because I had prepared a very excellent instrument for myself, I perceived (as I had not before, on account of the weakness of my previous instrument) that beside the planet there were three starlets, small indeed, but very bright."

Intrigued that these three starlets were on a line surrounding the planet, he observed them on following nights and found that they were changing positions with respect to Jupiter. On January 13, he saw a fourth starlet, to bolster his realization that he was observing moons traveling in orbits around Jupiter.

Galileo's observations, and the conclusion he drew from them that the Earth was not the center of the Universe, set into motion a tragic sequence of events. The year after Galileo's discovery of the moons of Jupiter, the first of several Catholic Church documents was filed secretly against him. In 1616, the publication of the Church's *Codex* specifically prohibited any teaching that the Sun, and not Earth, was in the center of heaven. Galileo wisely decided to wait until a more propitious time to defend the message of his telescope. With the election of Pope Urban VIII in 1623, Galileo thought that the time had come. The new pope seemed open to new ideas, and he spent hours walking and talking with Galileo. Nine years after the election of the new pope,

The author at work in his observatory. The telescopes pictured, left to right: a Meade ETX Maksutov; a 16-inch Dobsonian with 8-inch

in 1632, Galileo finally risked publishing his *Dialogue on the Great World Systems*.

The book was a catastrophe for Galileo. Somehow Galileo had misread one of the pope's objections, that the observations of his telescope did not seem to allow for the occur-

TELESCOPES

Meade telescope; a black Schmidt camera facing opposite direction; and a 6-inch reflector. (Photograph by Wendee Wallach-Levy)

rence of miracles. But the final straw was a character in the book, named Simplicio, who virtually ridiculed the pope's ideas. Enraged at the perceived insult from Simplicio, Urban placed the scientist at the mercy of the Office of the Inquisition. The aging and sick telescopist arrived in Rome in

February 1633, and two months later his trial began. Galileo was forced publicly to recant his observations that the Sun is the center. That December, he sadly returned to his home in Arcetri, near Florence, where he spent the last 8 years of his life under house arrest.

One bright spot during this tragic time was a visit from the poet John Milton in September of 1638. Milton was moved almost beyond words by this visit. In 1667, Milton published *Paradise Lost*. In Book V, he paid Galileo a great honor by mentioning his name:

> From hence—no cloud or, to obstruct his sight,
> Star interposed, however small—he sees,
> Not unconform to other shining globes,
> Earth, and the Garden of God, with cedars crowned
> Above all hills; as when by night the glass
> Of Galileo, less assured, observes
> Imagined lands and regions in the Moon;
> Or pilot from amidst the Cyclades
> Delos or Samos first appearing kens,
> A cloudy spot.

If Galileo was doomed, the course of the telescope he had set in motion was not. Galileo would never know how important a tool it would become centuries later. He would never know that a giant version of his telescope would orbit the Earth, and that a spacecraft named after him would observe a comet crashing into Jupiter, and then visit the moons he discovered.

Bigger and Better Telescopes

In Europe, telescopes quickly grew in size and popularity after Galileo. In 1656, Christiaan Huygens (who a year later would invent the first pendulum clock), used a telescope to discover that Saturn is circled by a ring. In 1664 Robert

Hooke used another instrument to make what appears to be the first observation of the famous Great Red Spot on Jupiter. As the apertures of telescope lenses grew larger, they also grew longer. Around 1670 Johannes Hevelius of Danzig, Poland, on the Baltic Sea built a stunning telescope 150 feet long. The "tube" was an open framework affair with a series of circular wooden "stops" to keep stray light from entering the eyepiece. Operated with the help of several assistants, the long tube was raised or lowered on a tall flagpole-like mast, and to keep the objective approximately aligned with the eyepiece 150 feet away, Hevelius used cords of adjustable tension. Since the telescope was unwieldy and would sway in the slightest breeze, it was difficult to use except on the calmest, clearest nights.

The year 1672 was a memorable one for telescopes. The first milestone occurred when 29-year-old Isaac Newton presented to the Royal Society a design for a telescope that replaced the objective lens with a mirror. The basic problem with using a *reflector* telescope is that when trying to look at objects in the telescope's mirror, the observer's head blocks the light from those very objects! Newton got around this by inserting a small flat mirror at a 45 degree angle near the top of the tube, to deflect light, and the observer's head, away from the main path. Newton's solution to that problem was so elegant that it has survived almost four centuries and remains one of the most popular telescope types today.

In May that same year came the announcement of a second reflector design by Guillaume Cassegrain, a royal sculptor whose many busts of Louis XIV still stand in the gardens at Versailles. Cassegrain's design offered some interesting differences from Newton's. Instead of using a flat secondary mirror to deflect the light, Cassegrain substituted a convex mirror that extended the light path instead of just bending it. The light was directed to return to the primary mirror and travel through a hole in its center, where the eyepiece was

COSMOLOGY 101

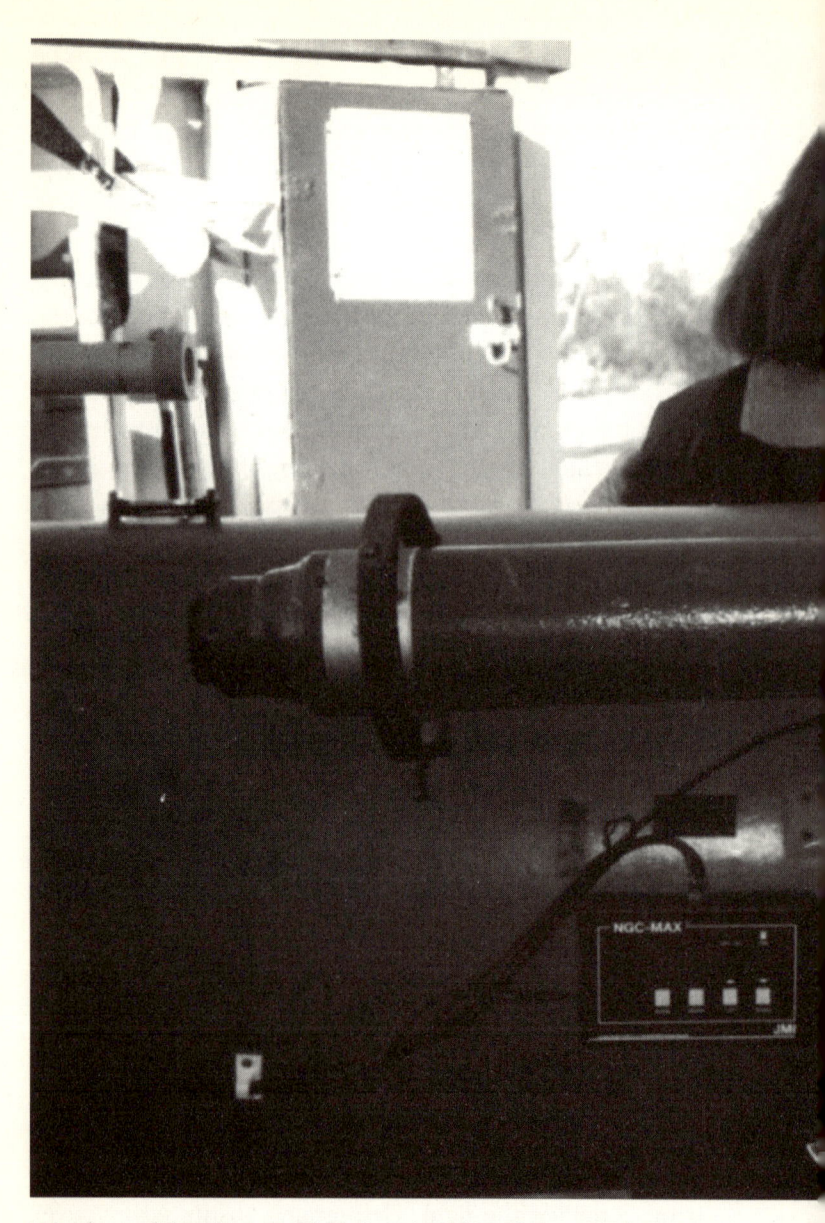

Wendee and David Levy with their 16-inch f/5 reflector. With this telescope, the author made seven comet discoveries between 1984 and 1994, and still uses it in his search for comets.

TELESCOPES

located. This new type of telescope was a variation of an earlier 1663 design by James Gregory, a Scottish mathematician.

Seven years after Newton and Cassegrain's achievements, on the morning of November 14, 1680, the German astronomer Gottfried Kirsch made the first discovery of a comet using a telescope (Yeomans, 1991). It was obvious that in Europe, telescope technology was increasing dramatically. But the rest of the world was slow to catch up. In early eighteenth century India, for example, Sawai Jai Singh, one of India's rulers, erected several beautiful structures of masonry and stone designed for a program of positional observations of the Moon and planets. Although he did own a telescope, he was apparently unimpressed with its ability to make measurements more accurately than he could with the unaided eye and his very stable masonry structures (Sharma, 1998).

Herschel Discovers a New World
By 1700, telescopes were common enough that astronomers throughout Europe were becoming familiar with them. Charles Messier was the first to use a telescope in the systematic search for comets, discovering his first in 1759. That year a musician's son, William Herschel, was adjusting to his move from Germany to England. Two years earlier, he had made the move to bolster his dream to become famous in music, and to raise funds for it, he frequently rode across the fields in typically damp and stormy British weather giving recitals. An example of the determination of this impoverished musician: In 1760 he was about to complete a trip to Genoa, Italy. To raise funds for his return to England, he gave a recital using a harp and two horns. He played the harp with one hand, held one horn with his other hand, and had a second horn slung over his shoulder. The Genoans enjoyed the recital and Herschel was able to make his way back home.

In May of 1773, Herschel bought an astronomy book, and

within a few weeks his passion for music suddenly transformed itself into one for the heavens. He built a four-foot long tube and mounted a lens in it. While observing from his backyard on the evening of March 13, 1781, Herschel discovered the first new planet in historic times. The discovery of Uranus made him famous, and the following year George III appointed him private astronomer to the King. By the end of that decade he had made a 20-foot long reflector, and soon after that he made one double that length. This great telescope had a focal length of 40 feet and a 4-foot wide mirror. Since the telescope was unwieldy it spent more time under repair than under the stars. Extremely difficult to use, the telescope's eyepiece could be reached only by standing on a high platform and shouting orders to the assistant, who almost always was his sister Caroline. An excellent astronomer who discovered eight comets on her own, she did like assisting her brother. But one night a large hook from the moving telescope struck and injured her ankle. By the end of his life, Herschel had completed more than two thousand telescopes. With his 40-foot-long reflector, Herschel detected the sixth and seventh satellites of Saturn, Enceladus and Mimas. With his 20- and 40-foot-long reflectors, Herschel catalogued over 800 double stars, and almost 2,500 nebulae.

A Leviathan of an Irish Telescope
The nineteenth century's biggest telescope was the life's dream of William Parsons, a man who was born into a sizable fortune and enjoyed a serious interest in astronomy. With the passing of his father in 1841, Parsons became Third Earl of Rosse and inherited his family's estate at Birr Castle in Parsonstown, Ireland. The next year Rosse began construction of what might be the most incredible telescope mounting ever built, two parallel piers of solid brick some 72 feet long and 56 feet high. Between them was slung a 72-inch diameter

reflector with which he succeeded in determining, for the first time, the spiral structure of many distant nebulae (King 1955). He also discovered the nature of many clouds of gas in our own galaxy; including a small planetary nebula with protruding rays that made it look like a dim version of Saturn with its rings.

George Ellery Hale

With the dismantling of the Rosse telescope in 1908, the world's largest operating telescope was the forty-inch diameter refractor at Yerkes Observatory. But telescopes were quickly getting bigger, thanks largely to the energy and drive of one man, George Ellery Hale, a solar astronomer and telescope fund-raiser of the highest order.

Early in the 20th century, Hale persuaded Charles Yerkes, a wealthy businessman, to fund a 40-inch diameter refractor on Wisconsin's William's Bay. This beautiful telescope is still the world's largest operating refractor. As his dream turned to bigger telescopes, he later arranged funding for a 60-inch diameter machine. He persuaded a businessman named J. D. Hooker to fund, with some additional endowment from the Carnegie Institution, the construction of a 100-inch reflector on California's Mount Wilson. But the achievement for which he is justly most famous for is the 200-inch reflector on Palomar Mountain. Although the great telescope was almost complete by 1939, the Second World War interrupted the work and the telescope did not see "first light" on the stars until June 1948.

The 1950s saw the origins of the National Observatory for the United States on Kitt Peak, southwest of Tucson, Arizona, as well as the development of a completely new kind of telescope for observing in radio wavelengths. These big dishes of radio telescopes helped define the structure of our galaxy. The next decades beheld a proliferation of large telescopes, especially in the southern hemisphere, as well as large numbers of radio telescopes. The 1980s saw the first small telescopes in space that observed at different wavelengths of light—optical, infrared, ultraviolet, X-ray, and gamma-ray, that culminated with the 1990 launch and 1993 repair of the Hubble Space Telescope.

The Hubble Space Telescope
Using a telescope on Earth is like looking up from the bottom of a lake. The atmosphere is a turbulent place, and the larger a telescope is, the more sensitive it is to defects in "seeing" caused by the flow of air in the atmosphere. Since the dawn of space flight, astronomers have dreamed of launching a large telescope into orbit. The successes of small space telescopes like the International Ultraviolet Explorer (IUE) only whetted their appetites.

Finally, in 1990, a space shuttle launched the Hubble Space Telescope into orbit about the Earth. During testing, the telescope consistently produced star images that were astigmatic. A few weeks after launch, astronomers learned that although the 90-inch primary mirror was ground to a precise tolerance, it was the *wrong* tolerance. The flaw was at first seen as the biggest catastrophe in the history of telescopes. Trying to recover from it, astronomers made the most of what they had by using the telescope—still better than any ground-based instrument—intensively during its first three years. It clearly recorded Pluto's moon, the quadruple quasar of Einstein's cross, and a host of other new things, until, in December 1993, all was ready for a repair mission to fit the Hubble with corrective lenses.

Launched by the space shuttle *Endeavour* in December of 1993, a crew of NASA astronauts spent several days in orbit working with the telescope. They knew that the future of astronomy in the ensuing decade was riding on the success of their mission, and after installing and positioning the corrective lenses during several difficult space walks, they returned to Earth and let the telescope get on with its work. The repair of the Hubble Space Telescope opened the door to the most stunning astronomical images ever seen. Just a month later, the telescope returned an image of the disrupted comet Shoemaker-Levy 9, and half a year later HST led the world's pack of scopes in recording the battered giant planet Jupiter as these same cometary fragments struck over a period of six days. Magnificent images of the Eagle Nebula captured by Hubble's "Deep Space Field" recorded galaxies farther off in space and hence further back in time than we had ever seen. If Galileo Galilei could return for one more visit, he would be thrilled beyond words to see how far his dream had progressed.

Types of Modern Telescopes

REFRACTOR, a telescope with an objective lens pointed at the sky and an eyepiece facing the observer. The design of Galileo's telescope involves light from a star entering the lens and converging to a focus at the eyepiece.

REFLECTOR, a telescope with a mirror to gather light and form an image. In a Newtonian reflector, light from a star strikes the mirror, which reflects it back up the tube to a small, flat secondary mirror that directs the light out the side of the tube, near the top, to an eyepiece. A Schmidt-Cassegrain reflector is modern, compact telescope that folds the incoming light. Starlight first hits a correcting lens, then the main mirror, then back to a secondary mirror, and finally back through a hole in the main mirror to the eyepiece, which sits beneath the mirror.

SCHMIDT CAMERA: Light passes through a correcting lens and then reflects off a larger primary mirror. A photographic plate or film, set between the mirror and lens, records the exposure.

REFERENCES

Bell, Louis. *The Telescope*. New York: McGraw-Hill, 1922, 2.

de Santillana, Giorgio. *The Crime of Galileo*. Chicago: University of Chicago Press, 1955.

Galilei, Galileo. "Discoveries and Opinions." Trans. Stillmann Drake. *Discoveries and Opinions of Galileo*. Ed. . n.p.: n.p., n.d. 287.

King, Henry. *The History of the Telescope*. 1955; rpt. New York: Dover, 1979.

Milton, John. "Paradise Lost." *Paradise Lost and Selected Poetry and Prose*. Ed. H. Northrop Frye. New York: Holt, Rinehart, and Winston, 1951, 1967. 115.

Sharma, Virendra N. *Sawai Jai Singh and His Observatories*. Jaipur, India: Publication Scheme, 1997, 76-77.

Sobel, Dava. *Letters to Father: Suor Maria Celeste to Galileo, 1623-1633*. New York: Walker and Co., 2001, 323-325.

Symmons. *The Life of John Milton*. London, 1822, 1970, 82-83..

Tennyson, Alfred. "In Memoriam XXI, 17-20." 1850. *Victorian Poetry and Poetics*. Ed. Houghton and Stange 51.

Yeomans, Donald K. *Comets: A Chronological History of Observation, Science, Myth, and Folklore*. New York: Wiley, 1991, 422.

CHAPTER 21

Using Your Telescope

"Tomorrow night,"—so wrote their chief—"we try
Our great new telescope, the hundred-inch.
Your Milton's optic tube has grown in power
Since Galileo, famous, blind, and old,
Talked with him, in that prison, of the sky.
We creep to power by inches. Europe trusts
Her 'giant forty' still. Even to-night
Our own old sixty has its work to do;
And now our hundred-inch ... I hardly dare
To think what this new muzzle of ours may find."

—Alfred Noyes

Alfred Noyes's poem "Watchers of the Sky," was a response to the poet's personal experience of being at the 100-inch telescope on Mount Wilson on its opening night in 1918. It refers to Milton's meeting with Galileo, to the earlier telescopes including the sixty-inch that now, as on that night in 1918, still has work to do. The telescope that rests covered in your closet also has its work to do; it waits a clear night to go out and help you discover the night sky. Go out in the evening, find your bearings on Observatory Earth, and meet this family. On your first night

USING YOUR TELESCOPE

out, it is enough just to look up and gaze at this miracle of a Universe we live in.

Aspects of the Sky: Observing from a Moving Planet

When looking up, the first thing to be aware of is that Earth, our home and our place of observation, is a huge ball of rock moving around the Sun through space. Planet Earth offers a stable platform from which we can observe and understand the rest of the Universe. But as we look up at the sky, what evidence do we have that we *are* observing from a moving ball, and not, as our ancestors thought, a flat surface supported by four elephants standing on the back of a turtle?

THE EARTH IS A SPHERE: We have at least three clear lines of evidence to this statement. The most common is to look for a sailing ship moving out to sea. If the Earth were flat, you would see the ship diminish in size until it became too small and disappeared. Because the Earth is curved, however, you see the ship appear to sink: first the hull, then the lower sails, and finally the upper mast vanish into the distant sea. Of course, looking across thirty to fifty miles of ocean will work only if the atmosphere is very steady, otherwise the atmosphere can play tricks on what you see.

An eclipse of the Moon provides another clue. The eclipse occurs when the shadow of the Earth is projected against the Moon. That shadow is clearly round, indicating that the Earth, its source, is spherical.

EARTH'S ROTATION: The movement of the Sun and stars from east to west indicates that the Earth is spinning in the opposite direction, from west to east. We see this effect any time we look toward the sky, and we plan our daily routines around this rotation of the Earth.

On your first night of observing, the Moon might be in the sky, and on succeeding nights it will appear in a different spot of the sky, east of where it was the night before. You can see this change both in the position of the Moon among the stars and in the changing times of moonrise or moonset, and it offers evidence that the Moon is orbiting around the Earth. The gravity of the orbiting Moon also affects the rise and fall of ocean waters twice each day, another line of evidence that Earth is rotating. Moreover, each day high and low tides occur about an hour later. This effect is due to the Moon's changing position as it orbits the Earth.

EARTH'S TILT AND ORBIT: Changing Seasons offer dramatic evidence both that Earth is a world orbiting the Sun, and that it is tilted on its axis as it does so. Because of this tilt, parts of the Earth away from the equator are tipped toward the Sun during half of each orbit and away during the other half. Because the Earth's northern hemisphere tilts away from the Sun in January, it experiences winter at that time. The southern hemisphere, which tilts toward the Sun at that time, has summer.

As you continue observing week after week, you will notice that although the hour is the same, the view has changed and stars that were in one part of the sky are now in another. From night to night the change is slight, as the stars rise about four minutes earlier. But that quickly adds up—a half hour per week, two hours with each passing month. The change happens because we look at the sky from a different perspective as the Earth orbits the Sun.

During your night out you may very well see a "shooting star." These flashes are not stars falling out of the sky but small particles of dust, mostly debris from comets, that encounter Earth's atmosphere. In its annual journey, Earth encounters several swarms of cometary dust. The richest occur on August 11 and December 12. At these times you can see meteors as often as once every few minutes.

It is easy to see the planets move slowly through the sky

during a season of observing. This movement is due to the planets orbiting the Sun. As the Earth overtakes the slower moving distant planets, they appear to slow down, reverse direction for a few months, and then resume their eastward trek. The effect is similar to passing a slower car on a highway. As we overtake it, the car appears to move backwards relative to us. This retrograde motion is evidence that the Earth is also moving rapidly along a highway in space.

Our Galaxy in Space
The night sky is studded with stars; in fact, on a clear night with no Moon, away from city lights, you should be able to spot some 3000 twinkling stars. The poetic twinkling of stars is the effect that inspired Jane Taylor's poem that begins "Twinkle, twinkle, little star." Written in 1806, this is the most famous star poem in the English language. Although light from distant stars has been traveling to us for years, or centuries, it travels on essentially a straight line until the last millisecond. But as each star's point source of light encounters the winds in Earth's atmosphere, it is bent. The result is twinkling.

Most of the stars we see in the night sky are the closest and brightest members of our galaxy, which we call the Milky Way. We can see the more distant members of our galaxy huddled in a structure that circles the sky—this is the Milky Way.

Imagine what our galaxy would have looked like five billion years ago. Although it probably had much the same shape as it has now, its stars were different. Unlike the many reddish suns we see now, most of the galaxy's early stars were big, hot, and blue. These stars lived hard and fast, each one burning out within a few million years and then losing its life in the violent outburst of a supernova. As these fast-track stars exploded, they spread their elements, including carbon, far and wide throughout the space around them.

Although these massive suns are not as common now as they were when the galaxy was young, there still are a few. The constellation of Orion, the Hunter, is riddled with them. The three stars in the Hunter's belt, as well as the bright star Rigel in his southwest corner, are good examples of massive, fastburning stars.

Long ago, another of these massive stars blew up and spread its remains into a nearby cloud of hydrogen particles. This cloud was very large and very cold, its particles hovering at several hundred degrees below zero. If you can see the Milky Way on this, your first night out, you might notice some examples of large, dark clouds in space. The most obvious one, the Great Rift (chapter 10), spans much of the length of the Northern Milky Way, dividing its light into two bands stretching from the constellations of Cygnus, the swan, all the way south toward Sagittarius, the archer.

As we have also seen in chapter 10, if it weren't for these dark clouds scattered through our galaxy, our Milky Way would be far brighter than we see it tonight. In fact, if there were no intervening dark clouds, the galaxy's central bulge would possibly be brighter than anything in the sky except the Moon.

Watching as New Suns are Born
After the supernova exploded billions of years ago, the giant cloud became infused with organic materials from the supernova. The cloud divided, each portion beginning to rotate around its center. Over a considerable time, the center grew quite hot. About 4.6 billion years ago, the center of the cloud was so hot it was shining. Continuing to enlarge until it reached a critical mass, one day it began to convert hydrogen to helium in a process we call nuclear fusion.

Although there is no solid evidence that we have seen a star begin nuclear fusion, there are many places in our galaxy where stars are about to be born. The most famous of these spots is the center of the Eagle Nebula. In 1996 the Hubble Space Telescope photographed its "pillars of creation," a symphony of bright and dark clouds shrouding the formation of a star and planets. Although we cannot see the pillars through a small telescope, we can see many spots in the sky, called bright and dark nebulae, where star formation is in progress just as it was with our system so long ago.

Observing the Planets

As our Sun was being formed out of its giant cloud, the material around the cloud's center coalesced into small proto-worlds several miles across as particles gently collided with each other. These small bodies, in turn, collided with one another, making larger bodies. In this manner, nine worlds were built over a period of several hundred million years.

On your first night out, you may be treated to a view of one or more planets. Venus and Jupiter are by far the brightest, and while Venus, if visible at all, can be seen only in the hours after sunset or before sunrise, Jupiter can lie almost anywhere, at times staying in the sky all night. Through a small telescope, Jupiter's moons shine like bright stars near their host planet, and Venus often shows phases.

Although Venus is much brighter than Jupiter, it is by far the smaller of the two. Venus is about the same size as the Earth, but Jupiter is so large that were it opened at the top, some 1300 Earths could pour in. When you look at Jupiter, remember that we owe that mighty planet a lot. Were it not for its rapid growth so long ago, we would probably not be alive here on Earth. Jupiter, after all, is the vacuum cleaner of the solar system. In the solar system's youth, comets were hurtling through the system, crashing frequently into planets like Earth. However, as Jupiter grew in mass, more and more comets were attracted by its gravity. Some of these comets collided with Jupiter, and the giant planet hurled many others out of harm's reach, thus sparing the Earth further bombardment. Comet collisions are rare today, but, as we have seen with Shoemaker-Levy 9, not totally absent.

Although fainter than Jupiter, Saturn is unique. Its rings are beautiful to see. And although it is not always visible in the sky, at certain times you might be fortunate enough to spot Mars, the red planet. You might think of how Earth relates to other worlds. While Earth somehow became an ideal habitat for life, Mars, smaller and colder, lost its atmosphere and its water while it was young, and Venus became so hot that no life would be possible on its broiling surface.

Each one of the three thousand stars you might see on a dark, clear night, is the center of a solar system. Each star was formed out of a giant molecular cloud, and as each star condensed, a system of worlds probably formed around it. How would the night sky look from one of these stars? If we could travel, for example, the 35 light years to the star Beta Geminorum, we would see a sky quite different from ours. The charts in this chapter compare Orion, one of the most famous constellations, as it appears here and as it would appear from that distant spot in the sky. Other constellations would look different too, although the more distant outposts of the Milky Way would appear much the same as they do from the Earth.

We've come a long way for only a night or so of observing. But have you ever imagined what a treasure awaits you as you walk outdoors one evening and look out at the stars?

REFERENCES

Noyes, Alfred. *The Torchbearers: Watchers of the Sky*. n.p.: Frederick A. Stokes Co., 1922.

CHAPTER 22

Notes on Western U.S. Observatories

Our hero felt the wind's first lash.
He felt the distant lightning's flash.
"My dear!" he cried! "I shall return
Some other night with lips that burn.
Right now I've got to dash for home
For I forgot to close the dome."
—Dorothy Peltier, (circa 1920)

The excitement of opening an observatory dome at nightfall is shared by amateur and professional astronomers alike. The experience is like those few moments in a theater when the audience awaits the curtain rising. Slowly, majestically, the dome shutters move apart, revealing the darkening sky whose brightest stars are already tuning up for the overture. Although I have seen hundreds of these observatory nights, I'm still mesmerized by the wonder of the moment the dome opens.

When Harlow Shapley and his new wife headed west in

1915, they were uncovering a new frontier and an open observatory dome. Sitting on a train and studying the light curves of stars that change in brightness, they planned a lifetime of observing the night sky from the newly opened 100-inch reflector telescope atop Mount Wilson near Los Angeles.

It didn't quite turn out that way. Within a few short years, Shapley had used the great Mount Wilson telescope—then the largest in the world—to prove that the Sun was not in the center of our galaxy but instead some thirty thousand light years out in its boondocks. After that, the Shapleys returned east, where Harlow became director of the Harvard College Observatory.

Shapley's trip to Mount Wilson was one of the early treks, but hardly the first. In the 1890s Percival Lowell, a member of one of Boston's most famous families, founded an observatory in Flagstaff, Arizona so that he could search for evidence of life on Mars. In 1930, Clyde Tombaugh used that observatory to accomplish Lowell's other dream of discovering a new planet.

In the past century many people who loved the stars headed for the prevailing clear skies of the great American west, and while they are gone, their observatories thrive. Here are the results of these efforts—a group of fascinating, mostly mountaintop havens where astronomers study the sky. These places are worthy destinations for a vacation trip.

Lowell Observatory
This is the place that Percival Lowell made famous through his Mars and Planet X studies, and the observatory retains an important feeling of its own history. Some of its best early accomplish included astronomer Vesto M. Slipher's discovery of the red shift in 1903, Clyde Tombaugh's discovery of Pluto in 1930, and the 20-year Proper Motion Survey that began in the 1950s. More recently, LONEOS, the Lowell

Observatory Near Earth Asteroid Survey, searches the sky for asteroids that might collide with Earth.

Located on Mars Hill in Flagstaff, Arizona, the observatory has one of the best public education programs of any professional research institution. When I toured it in 1967, our tour group was taken around by Robert Burnham, a staff member who had discovered several comets. Even today tours are usually conducted by staff scientists.

http://www.lowell.edu/

Mount Wilson
The story of Mount Wilson actually begins at Yerkes, near Chicago, where George Ellery Hale dreamed of taking advantage of the clear skies of the American West. Its 60-inch reflector has one of the finest mirrors ever produced, and its 100 inch Hooker reflector was the largest telescope in the world from 1918 to 1948.

Located in the Angeles National Forest east of Pasadena, the observatory is reached through a spectacular mountain drive and is well worth visiting. The Mount Wilson Observatory Association is affiliated with the observatory and studies its rich history.

MOUNT WILSON OBSERVATORY =
    http://www.mtwilson.edu/
MOUNT WILSON OBSERVATORY ASSOCIATION =
    http://www.mwoa.org/

Lick Observatory
James Lick wanted a monument to himself, perhaps a pyramid in San Francisco Bay. Instead, he was persuaded to spend his fortune on an observatory near San Jose at the southern end of the Bay. It is one of the centers for research on extrasolar planets. The observatory is east of San Jose, California.

http://www.ucolick.org/

Palomar Observatory

The home of the 200-inch Hale reflector, Palomar Observatory is one of the most majestic astronomical sites in the world. Russell Porter, an amateur mirror grinder from Vermont, completed the optics for the 18-inch Schmidt camera that was later used in almost 50 comet discoveries, including my co-discovery of Comet Shoemaker-Levy 9. The observatory also has a 48-inch Schmidt camera and a 60-inch reflector.

The observatory is about a forty-five-minute drive from Escondido, California.

http://www.astro.caltech.edu/observatories/palomar/

McDonald Observatory

The story of the observatory in the Davis Mountains of west Texas is a fascinating one that begins with Georges van Biesbroeck at Yerkes Observatory. "Van B's" dream was to build an observatory in one of the truly darkest sites in the world, and 70 years later, McDonald still has that distinction. In the 1970s, a depressed employee attacked the mirror of the observatory's 107-inch telescope, shooting it with a rifle and then trying to carve it with an axe. Miraculously, the thick mirror sustained almost no damage at all!

As a result of the Apollo Moon landings, one of the observatory's best known projects was to shoot laser beams off the Apollo reflectors. This data allowed calculations of the Moon's distance to Earth in an accuracy of a fraction of an inch.

McDonald is some distance east from El Paso, Texas along Interstate 10, or west from Austin.

http://www.as.utexas.edu/mcdonald/mcdonald.html

Kitt Peak and Steward Observatories

In the mid 1950s, a group of astronomers were worried that aside from those who worked at the great observatories,

NOTES ON WESTERN U.S. OBSERVATORIES

astronomers throughout the country could not readily have access to large telescopes for their research. Accordingly the National Science Foundation decided to fund a national observatory atop Kitt Peak near Tucson. Since the land was part of an Indian reservation whose officials called the astronomers "men with the long eyes," it was important to give them a view through an existing telescope. That single observing session set the stage for the observatory that thrives on a mountaintop southwest of Tucson, Arizona.

KITT PEAK VISITOR CENTER =
http://www.noao.edu/outreach/kpoutreach.html
KITT PEAK NATIONAL OBSERVATORY =
http://www.noao.edu/kpno/
STEWARD OBSERVATORY =
http://www.as.arizona.edu/steward/

The MMT and the Smithsonian Astrophysical Observatory

In 1979 a completely new kind of telescope was opened atop Mount Hopkins, south of Tucson. Its prime telescope used six mirrors instead of one to work together as a multiple mirror telescope within a single tube assembly. In 2000 the mirrors were replaced with a single, large mirror. This truly unique telescope is located near Amado, Arizona, south of Tucson. Tours available by appointment.

http://sculptor.as.arizona.edu/foltz/www/mmt.html

The Very Large Array

Poised against the sky in mountains near Socorro, New Mexico, stands the multiple dishes of the National Radio Astronomy Observatory's Very Large Array. The outerworldliness of these huge radio antennae is unique in the annals of telescope building, and the telescopes have been used in two major motion pictures, Arthur C. Clarke's *2010* and Carl Sagan's *Contact*. These telescopes do not look into space, they listen; and all the energy that has been detected by these telescopes, plus every other radio telescope on Earth, is less than that of a single snowflake.

http://www.aoc.nrao.edu/vla/html/VLAhome.shtml

Mount Graham International Observatory

In the 1980s, several universities got together to build a series of large telescopes high on Mount Graham, in southeastern Arizona. After a protracted battle with environmentalists, who believed that the Mount Graham red squirrel could be rendered extinct by the construction, the observatory was nearing completion in the early years of the new millennium. The Vatican Observatory, based in Italy, built one of the telescopes.

Although this facility is not currently open to the public, in nearby Safford, Arizona, there is a gem of a science center called Discovery Park. This mostly outdoor center showcases all that is beautiful about the nature and science of southern Arizona, and highlights the interaction of Arizona's people with the desert's delicate ecology, and with its pristine night sky. There is also a 20-inch telescope for public viewing of the sky.

MOUNT GRAHAM =
http://medusa.as.arizona.edu/graham/graham.html
VATICAN OBSERVATORY =
http://clavius.as.arizona.edu/vo/
DISCOVERY PARK =
http://www.discoverypark.com/main.html

Observatories in Southern New Mexico

The New Mexico State University founded an observatory during the 1960s, in the backyard of one of its graduate students, to study the planets in greater detail than had ever been attempted before. Under the leadership of Clyde Tombaugh, discoverer of Pluto, this observatory, suitably relocated atop Tortugas Hill near Las Cruces, succeeded in its mission. Other related observatories, like the National Solar Observatory in Sunspot, and the nearby Apache point observatory, have also opened in recent years.

APACHE POINT OBSERVATORY =
  http://www.apo.nmsu.edu/
SACRAMENTO PEAK OBSERVATORY =
  http://www.sunspot.noao.edu/sunspot/sp_index.html
LAS CRUCES-BASED OBSERVATORIES =
  http://astro.nmsu.edu/observatories.html

LINEAR's Search for Comets

In the central part of New Mexico lie two small domes that house completely automated telescopes that search the sky for comets as part of the Lincoln Laboratory Near-Earth Asteroid Research. In the last few years there have been more comets discovered by LINEAR than any found from an observatory based on the ground.

  http://www.ll.mit.edu/LINEAR/

Following Deep Space Missions

In the Mojave Desert not far from Barstow, California, stands a tall antenna that is not really an observatory unto itself. But it and two other antennae, located around the world, catch signals sent by distant spacecraft, like *Galileo* and *Cassini*, as they speed by the distant worlds of the outer solar system. This is NASA's Deep Space Network, its ear to the edge of the solar system.

  http://gts.gdscc.nasa.gov/pages/dsnintro.htm

REFERENCES
Peltier, Leslie C. *Starlight Nights: The Advntures of a Star Gazer*.
  1965: rpt. Cambridge: Sky Publishing, 1999.

RICHARD RUBIN

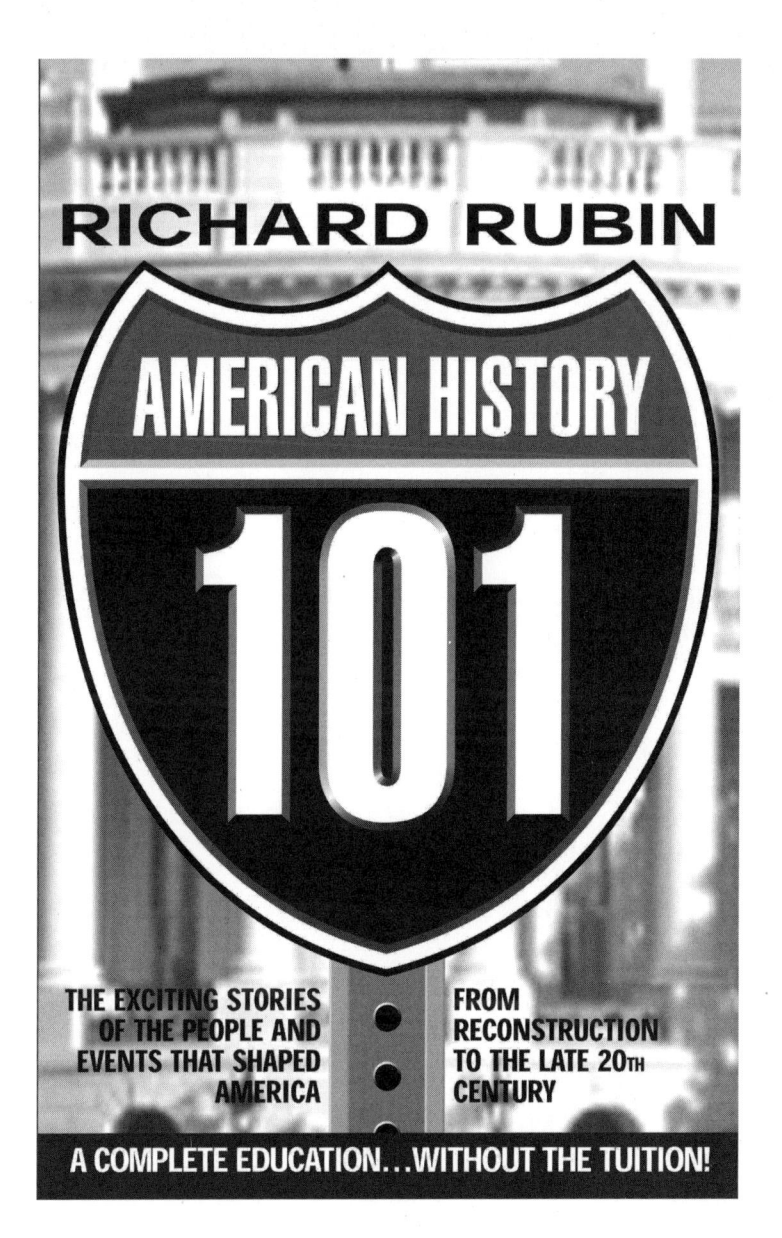

AMERICAN HISTORY 101

THE EXCITING STORIES OF THE PEOPLE AND EVENTS THAT SHAPED AMERICA

FROM RECONSTRUCTION TO THE LATE 20TH CENTURY

A COMPLETE EDUCATION…WITHOUT THE TUITION!

AT BETTER BOOKSTORES NOW

ROBERT SILVERBERG

SCIENCE FICTION 101

WHERE TO START READING AND WRITING SCIENCE FICTION

"An excellent introduction to the most important roots of modern SF."
—The Washington Post

ORIGINALLY PUBLISHED AS ROBERT SILVERBERG'S WORLDS OF WONDER

AT BETTER BOOKSTORES NOW